What Your Colleagues Are Saying...

Seamlessly bridging the gap between foundational concepts and advanced insights, *Let's All Teach Computer Science!* Offers a refreshing perspective for seasoned educators, while providing clear guidance for beginners. A must-read to elevate computer science instruction across all levels.

Aria Azizi
Senior Education Program Manager, Minecraft Education
Redmond, WA

Reading this book felt more like sharing great ideas with a colleague in a really good professional development (PD) session than "reading a book." The conversational tone appeals clearly to the newer computer science teacher, and offers plenty of rich insights and perspective I picked up along the way through this book for my own work as a veteran computer science (CS) teacher, too. This is the kind of book that should be in the professional library of every current or aspiring CS teacher.

Steven Svetlik
Computer Science Teacher, Ridgewood High School
Founder and Executive Director, CS4IL
Norridge, IL

Prottsman's book is a thoughtful exploration of various ways educators of all disciplines can incorporate meaningful lessons using computer science. She focuses on tangible, achievable approaches and offers supportive guidance along the way. I particularly loved Prottsman's anecdotes, resources, and questions, which tie her guidance to real-world scenarios. This book helps envision how we can enhance students' lives by using tools in computer science!

Jen Fox
Engineer, Maker, Educator, Microsoft
Seattle, WA

This comprehensive guide has a wealth of insight and ideas that show how and why to incorporate computational literacy across the PK–12 curriculum, and equips teachers with the tools and strategies to foster computational skills in all students. It's the best way to invest in a future where technology is understood and can be used to do good.

Dianne O'Grady-Cunniff
Director, Maryland Center for Computing Education
Adelphi, MD

Let's **ALL** Teach
COMPUTER
SCIENCE!

Dedication

This is where I get to acknowledge how impossible it would have been for me to create this book on my own.

Namely, I want to thank my editor, Debbie Hardin, for being the ultimate debugger in this literary codebase—thanks for helping me stay on schedule and avoid syntax errors.

Thank you to Jake Baskin for writing the foreword to this book. I'm so glad that you were willing to lend your voice and provide us with the essential "Hello, World!" for this educational journey. I adore you and respect you like crazy.

To my amazing friends and colleagues who have spent hours and hours of their time chatting with me, sharing ideas, and reviewing the early versions of this book—thank you for being my real-life stack overflow.

And, to my family, who occasionally received the "404: Kiki Not Found" error during this project—thanks for staying on the page. I love you all.

I'd also like to send my sincerest appreciation to:

Maggie David
Becca Dovi
Jenna Garcia
Joanna Goode
Katie Hendrickson
Bill Marsland
Jacqueline Russell
Dan Schneider
Maria Sellers
Perry Shank
Holly Swartz
Shana White

Let's **ALL** Teach COMPUTER SCIENCE!

A Guide to Integrating Computer Science Into the K–12 Classroom

Kiki Prottsman

Foreword by Jake Baskin, Executive Director
of the Computer Science Teachers Association

CORWIN

For information:

Corwin
A SAGE Company
2455 Teller Road
Thousand Oaks, California 91320
(800) 233-9936
www.corwin.com

SAGE Publications Ltd.
1 Oliver's Yard
55 City Road
London, EC1Y 1SP
United Kingdom

SAGE Publications India Pvt. Ltd.
Unit No 323-333, Third Floor, F-Block
International Trade Tower Nehru Place
New Delh-110 019
India

SAGE Publications
Asia-Pacific Pte. Ltd.
18 Cross Street #10-10/11/12
China Square Central
Singapore 048423

Printed in the United States of America.

Paperback ISBN 978-1-0719-3152-3

Vice President and Editorial Director:
  Monica Eckman
Associate Director and Publisher,
  STEM: Erin Null
Acquisitions Editor: Debbie Hardin
Senior Editorial Assistant: Nyle
  De Leon
Production Editor: Tori Mirsadjadi
Copy Editor: Talia Greenberg
Typesetter: Integra
Cover Designer: Scott Van Atta
Marketing Manager: Margaret
  O'Connor

This book is printed on acid-free paper.

24 25 26 27 28 10 9 8 7 6 5 4 3 2 1

Contents

Part 2: A Deep Dive on Integrating Computer Science Into the Classroom   41

Part 3: Teaching Computer Science for Today and for the Future   77

For additional information and resources, visit the author's blog at **https://medium.com/geek-groupies**

Note From the Publisher: The author has provided web content that is available to you through QR (quick response) codes. To read a QR code, you must have a smartphone or tablet with a camera. We recommend that you download a QR code reader app that is made specifically for your phone or tablet brand.

Foreword

If you've purchased this book, you've already taken a crucial step toward teaching computer science, and you're already ahead of where I found myself during my initial days in the computer science classroom.

I came to teaching computer science in the Chicago Public Schools with more experience in computer science than teaching, and a ton of energy to bring something I am passionate about to my students. I was lucky to have a principal who created space for a full-time computer science teacher before it was the cool thing to do, a full slate of computer science classes to teach, and many challenges ahead of me. My first classroom was long and narrow, with rows of desks bolted to the ground and a collection of huge CRT monitors (never enough for each student, of course). If you remember those monitors, you know that they offered a lot of hiding spots for students, especially in the back rows. I'll never forget one evening toward the end of my second week of teaching. I was going computer by computer to update software, and when I got to the last row I found a pile of keys that had been popped off of the keyboards. In the center computer was a full keyboard, with a rather elaborate string of curse words spelled out on each row of keys. High school students have such creative minds that they apply in ... very interesting ways.

This early challenge was a vivid lesson in an essential truth: The crux of teaching computer science lies more in the "teaching" than the "computer science." If you're someone who's looking to bring computer science into your classroom, don't forget that you already have a wealth of knowledge that you bring to the field. In fact, at my day job with the Computer Science Teachers Association we do a biannual report on the landscape of computer science teachers, and half of all computer science teachers in the United States are new to teaching the subject (and experienced as teachers overall). I know it can feel daunting to start teaching new material, especially if it's content that you might be learning yourself, but you're not alone. And with Kiki Prottsman's *Let's All Teach Computer Science!*, you'll literally have in your hands the answers to most of your questions.

In fact (spoiler alert!), my absolute favorite question in this whole book appears early on.

Q: Exactly how much are we expected to know about CS before handing out laptops and tossing our students into the computer science pool?

A: About 50 minutes' worth of material more than what your students know.

With this book, you'll be well ahead of that 50-minute mark! Kiki has thought through the essential questions for getting started teaching computer science, and importantly done so in a way that will build your confidence and ability not only to leverage the many resources available for teaching computer science, but also to discern what might be the best fit for you and your students. The format makes for an easy read straight through and as a reference as you run into specific questions the night before (or in the middle of) your lessons integrating computer science.

Which leads me to my second favorite part of the book: Kiki's focus on supporting you in building up skills to create a curriculum that is right for your students. As you experiment with the many different resources outlined in this book you'll get a sense of their strengths and weaknesses, while at the same time you'll start to see what resonates with your students. No two classrooms are the same, and I strongly encourage you to start adapting computer science teaching resources as you gain confidence in your classroom. I say this not only because it will allow you to engage more students (it will!), but because it will build ownership over the material and confidence in your abilities.

There are many reasons why you would want to bring computer science into your classroom, from economic opportunity to civic engagement, and as you learn from your students, you'll discover new and more nuanced motivations for yourself. My hope for you, holding this book and embarking on this journey, is that you will embrace the creativity of the field, because at the end of the day through computer science you're giving your students the ability to create with some of the most powerful tools available, and I can't wait to see how they'll use those tools to shape the world.

—**Jake Baskin**, Executive Director,
Computer Science Teachers Association

Publisher's Acknowledgments

Corwin gratefully acknowledges the contributions of the following reviewers:

Charlotte Cheng
Director of Education, CodeCombat
San Mateo, CA

Rebecca Dovi
Chief Computer Science Advocate, CodeVA
Richmond, VA

Bill Marsland
Director of Education, Code Ninjas
Los Angeles, CA

Daniel Schneider
Principal Curriculum Content Developer, Code.org
Tucson, AZ

About the Author

Kiki Prottsman is an expert in computer science education and an advocate for equity and inclusion in STEM fields. With over 15 years of experience teaching and developing educational programs, Kiki has made significant contributions to the field of CSEd. She is also an author of several other books on computer science that have been widely praised for their innovative and engaging approach.

Kiki is currently Director of Education for Microsoft MakeCode, an organization that works to increase access to computer science in schools across the United States. In this role, she oversees the development of curriculum and resources as well as training programs for teachers. She also builds and manages partnerships with stakeholders worldwide.

Kiki speaks internationally on the subject of computer science. She is known for her engaging and practical approach to teaching and has been recognized with awards such as the Golden Halo Award for Best Education Campaign and Stevie's Female Innovator of the Year award.

In addition to her work in computer science education, Kiki is an outdoors enthusiast and enjoys hiking, kayaking, and camping, as well as customizing her 4 × 4 SUV to look like it belongs in a Marvel movie.

Introduction

I wrote this book for you.
Literally *you*.

I welcome you to participate in the process of building a safer and more united tech landscape for yourself and for the adults of tomorrow. I have had far too many experiences in my life where I felt like an interloper, which is a sure sign that those events were not as inclusive as they could have been. My goal with this book is to help you understand that you belong in this world of computer science education—and because of you, there will someday be an adult in the tech industry who is more knowledgeable, ethical, and aware because they had you in their life.

If you take a look at the Acknowledgments section, you'll notice that this book was not written based on my viewpoint alone. I've spoken to dozens of individuals about this content, including experts in computer science education, **equity** advocates, industry professionals, curriculum creators, and teachers who are new to the process of integration. I've performed a lifetime of research, exploration, and experiments in this space—so you can rest assured that the content presented here can give you everything you need to be considered an active part of the computer science education community.

To keep us on the same page, I will start by providing you with any background you'll need to digest the glorious gems of insight that I'm about to mine. It doesn't matter if you teach art to 5-year-olds, science to 10-year-olds, or math to 18-year-olds—my hope is that you will find a great deal of inspiration in the pages to follow, even if some of the material feels foreign at first.

As an artist-turned-computer scientist, I understand that mental barriers can pop up and turn even the most benign concepts into daunting hurdles. That same artist-to-computer scientist transition allowed me to see the beauty, creativity, and power that comes from incorporating **technology** into a traditional routine.

Computers are not soul-sucking machines that drain curiosity and independence from imagination. On the contrary! Computers are tools of innovation that allow the wielder to iterate through endless possibilities in less time than it would otherwise take to put initial thoughts on paper.

Using Common Vocabulary

As we walk along this path together, you may notice that I often use the terms *technology*, **computer science**, and **coding**. In education, these terms get conflated, but I'll ask that you keep in mind some important distinctions among the three.

In general, the word *technology* includes all things innovative. In the 15th century, the printing press was considered a huge technological innovation.

In the 18th century, pencils were the latest technology. Paper currency, steel, lightbulbs—all were technological advancements in their day, and all faced bitter opposition before they became commonplace. Within this book, I will use the word *technology* to refer to present-day computerized equipment and the various accouterments that surround it.

The term *computer science* (CS) means different things to different people. Within U.S. grade schools, it's often reduced to *coding*. While coding *is* a part of CS, it's only a fraction of a rich and complicated landscape. Computer science is a wide field of study that deals with all things related to the theory, design, and implementation of computer systems (including both **software** and **hardware**). At its core, computer science is the study of problem-solving using computers. To cover all bases, I'm going to lean on the definition adopted by the Computer Science Teachers Association (CSTA): "Computer science is the study of computers and algorithms, including their principles, their hardware and software designs, their implementation, and their impact on society."

You may already be familiar with the word *coding*. At a high level, coding is the art of telling computers what to do by writing instructions in a language they understand—however, as computer science expanded into elementary school, the definition of coding also expanded to include the creation of instructions for any entity ... whether it be using arrows on paper to direct a friend to stack cups in the correct order or using a series of claps and stomps to direct a teacher to provide exaggerated facial reactions. While some may consider this a "watering down" of the craft, I believe the new definition allows for age-appropriate lessons and spiraled learning, so that's the one I use.

While I'm distinguishing terms, let's look at *coding* versus **programming**. To the average person, coding and programming mean the same thing. There is a small subset of people who will happily want to correct me and point out that programming involves not only writing code but also designing the structure and logic of a program or software. Programming does, in fact, tend to include the creation of a blueprint for a **digital** system as well as the actual coding part, but I won't be making that distinction within these pages.

This book focuses mainly on the concept of coding in the classroom, but I *will* touch on other computer science ideas when the integrations feel appropriate.

What to Expect From This Book

Are you ready to become a fearless explorer of the digital domain, armed with knowledge, curiosity, and a touch of whimsy? If your answer is a resounding "Yes!" then keep reading and let the adventure begin!

I'd like to start by describing what you won't find in this book. As a former teacher, I promise that I won't be filling these pages full of propaganda from within the tech industry. I may work at Microsoft now, but I'm a firm believer that educational solutions should be born from research and experience, not deep pockets. To reflect this, I'll be spotlighting stories from my own life that highlight my positions and discoveries, allowing you to make up your own mind about how and why you adopt the strategies that I suggest. You will, however, mostly see activities and suggestions that utilize MakeCode Arcade, since that's the platform I have access to on the daily.

I won't be leaning on complex jargon or white-paper formality. I wrote this book in the same way I would have shared with you in person, meaning that you'll undoubtedly run into my famous simplified analogies and cringey humor along the way. I don't apologize!

What you *can* expect is a straightforward look into my many years of experience working with students in the field of computer science education, **computational thinking**, and coding. My suggestions have been distilled from decades of experience, including lots of trial and error and years of discussions with teachers of all experience levels.

If you need clarification on anything at any time, feel free to set the book aside for a moment and tag @KIKIvsIT in a tweet. I'll do my best to explain ideas and suggest additional resources.

We'll start by gaining a shared understanding of what CS Integration means for K–12, and why it's important. Afterwards, we'll peek into some of the things you'll need to know before adding CS into your classroom. Then, I'll reveal the techniques that I've used this past decade to create hundreds of lesson plans for teachers of all subjects and age ranges.

At the end of each chapter, I'll present a set of review questions that you can use to see if you've absorbed the core ideas. I encourage you to physically write your answers directly in the book so that when you get to the end, you can look back at the progress you've made along the way.

If you're running short on time, you're welcome to passively read the text and highlight ideas you want to come back to later. If you want to dive in more deeply, try connecting with our online community to discuss your plans (or points of confusion) so other teachers can benefit from the epiphanies that you've enjoyed. Education is a group activity, and together we can stop the gatekeeping that prevents our kids from discovering their true potential.

Locating Supplemental Materials

When it's time to gather your arsenal of resources, I'd love for you to think of yourself as a treasure hunter who is seeking out the best tools and materials to make your CS-fortified lessons shine. There are countless online platforms, interactive apps, and educational games available to engage your students and make learning fun. Don't be afraid to experiment and discover what works best for you and your classroom. Try searching for terms like "Coding for Kids" and "Computer science in the [*your subject here*] classroom" to see what comes up.

Also, remember that collaboration isn't just for kiddos! Lean on your fellow teachers for support. Form study groups, share lesson plans, and bounce ideas off of each other. Together, you'll create a safe haven where fears are conquered, and triumphs are celebrated.

In many cases, additional resources have surfaced during my research and interviews. These items are too good to miss, so I've gathered as many as I can and will connect them to this book via the online Appendix at https://medium.com/geek-groupies.

Are you ready to jump in? I know that I am!

PART 1

GETTING STARTED WITH COMPUTER SCIENCE INTEGRATION

We are capable of more than we know.

"There are no constraints on the human mind, no walls around the human spirit, no barriers to our progress except those we ourselves erect."

—Ronald Reagan

CHAPTER #1

The Importance of Integrating Computer Science Into K–12

Q: Can you tell me more about the end goal here? Not every student needs to go into tech after they graduate. Why is there such a push to make sure that every student learns computer science?

A: Such a great question! I think we can all agree that not every student will become an author or an athlete, either—but we recognize that reading, writing, and running provide advantages that individuals without those skills will never have. The same is true with future-ready skills like coding, data analysis, and IT. Whether or not someone winds up in the tech sector, they're very likely to need CS skills if they want any sort of innovative job in the years to come.

The title of this book likely means something different to you, depending on who you are and where you come from. If you teach social studies in the Pacific Northwest, where CS requirements are nearly non-existent, you may be hoping this book is full of simple tips and tricks for adding meaningful projects into your class as a way of enriching 21st century learning. If you teach biology in Memphis—where computer science requirements were recently added at high school, middle school, *and* elementary school levels—you may hope that this book is going to provide approachable guidance for bringing authentic computer science into core classes. On the other hand, if you teach mathematics in New Brunswick—where K–12 computer science requirements have been a thing since 2017—then you might be hoping to find something new and unique to shake up the lesson plans that you've been using for years.

I have tried to address all those hopes.

Within the context of this book, "integrating computer science into K–12" is distinct from the goal of adding additional CS classes to the school. While that's also a noble cause, my intention is to start the "Age of Computer Science" with a world full of teachers who recognize CS as a suite of tools and techniques that can be employed to solve any problem or enhance any task.

Does this mean that I think teachers need to add computer science into every lesson for every subject? I won't admit to that—but I will say I equate computer science to reading, writing, and drawing in that it is a skill that can be used each day to enhance topics and create deeper understanding of the world around us.

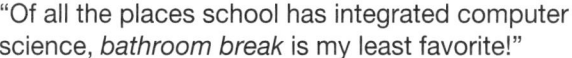

"Of all the places school has integrated computer science, *bathroom break* is my least favorite!"

Source: Kiki Prottsman 2024. Created using Midjourney and edited with graphic design programs.

The Marvelous Art of Computer Science Integration

Let's delve deeper into how this fusion between computer science and other subjects takes place. Picture a math class where students not only learn traditional equations but also explore the creation of step-by-step recipes to help peers solve similar problems in the future. Those are called **algorithms,** and those are fundamental to computer science.

In a history lesson, students might use digital tools to look at historical trends and compare the way prices changed in certain areas when hit by drought or tragedy. That's data analytics, another very popular piece of computer science.

Art, biology, choir … any class can look at the way computer science is affecting the industry and bring small, entertaining lessons into everyday education. While these additions are capable of being massive and complicated, they don't have to be complex for students to develop foundations for the future.

Keep in mind that integration isn't about overshadowing your core subject, but about making harmonious enhancements so students naturally absorb CS concepts and vocabulary. It's about selecting relevant computer science ideas that naturally fit into your existing curriculum. The goal is to make integration so seamless that the focus stays on your material, and students view CS as a multitool for future problems.

Integration should be customizable according to educator preferences and student needs. Someday, districts might start dictating exactly how and when each non-CS class should hit various computer science standards, but until that day comes, you get to choose how much digital pizzazz you want to add to your teaching repertoire. That freedom may bring some level of uncertainty, but a jam-packed toolkit will help make the process easier for you and your students.

You may find that you prefer a collection of one-off coding activities, allowing you to provide a minidigital adventure for your classroom. Or you might want an extensive, year-long journey where computer science becomes the trusted sidekick in all of your lessons, like Watson for Holmes. Either way, it's no mystery that computer science has the capability of enhancing subjects when introduced properly.

Why Integration?

Computer science is one of the fastest growing fields in any developed nation. CS exists within every field, from medicine to agriculture to theater—and jobs in computer science tend to pay disproportionately well when compared to other options that require little more than a bachelor's degree.[1]

Even if we have no interest in educating students to become professional computer scientists, the skills that are taught in CS classes have been shown to bolster students in relation to both executive function[2] and standardized test scores.[3] This fact amplifies my point that computer science education is more than a path to the tech sector. It's also a method for enhancing a student's educational growth by providing lifelong support in all of their scholastic adventures. That's why I promote it, and that's why I'm writing this book.

It should come as no surprise that much of the current computer science landscape is excessively academic. As a subject, it springs from one of the most complex and logic-filled systems to be developed by humankind, which exacerbates the idea that it's not to be tampered with, except by an exclusive braintrust. This myth creates an unnecessary elitism within the field, breeding a world where technology and programs fail to represent the people who use them.

Until recently, most children weren't introduced to computer science until college, and even then the subject was reserved for students who intentionally planned to study technology. This created a cycle of wealthy tech intellectuals who would then coach their children, making it nearly impossible for someone outside the industry to feel included as a first-year CS student.

To tackle this phenomenon, research shows that we should be reaching out to kids before they solidify their concept of what they can or can't be when they grow up. Unfortunately, this happens extraordinarily early and, according to Camp and Gürer,[4] access to computers and training in computer science should begin at preschool levels to give students from historically underrepresented groups the greatest chance against developing insecurities around their capabilities.

Even still, the school system is slow to change, and with budgets that are dwindling and teachers who are overworked, the chances of incorporating dedicated CS classes in K–5 across the world will remain low well into the next decade.

That's where you come in.

Together, you and I can work to integrate computer science and coding into classes that students are already taking so we can help them develop the reasoning, persistence, and holistic mindset that it takes to be truly resilient innovators.

Computer science also promotes problem-solving. When faced with difficulties, students educated in computer science can come prepared with a toolbox overflowing with logic, creativity, and critical thinking skills. With these practical gizmos, they are more likely to be able to dig themselves out of issues that initially appear to be over their head.

But wait, there's more! Integrating computer science into other classes also unleashes the art of collaboration. Your classroom can be full of students huddled together brainstorming ideas, debugging code, and making digital discoveries. The power of teamwork combined with the boundless possibilities of technology can create wonders that no single individual could accomplish alone.

The power of teamwork combined with the boundless possibilities of technology can create wonders that no single individual could accomplish alone.

As if that weren't enough, computer science also has a huge PR issue, and I believe CS integration can fix this.

Imagine how different it is to have your first introduction to computer science happen through the creative cajoling of an art or music teacher versus the mandatory requirements of a pedantic professor. Up until this point, most people who have studied computer science found the latter, but the former has so much more potential when it comes to reaching people who wouldn't generally self-select into the subject the way it is now.

Coders tend to be drawn to code and artists tend to be drawn to art. What if we could create more artists who liked to code and coders who liked to art? Imagine the effect that would have on the future when it comes to the people represented within the tech industry … and therefore the way that technological products represent the communities that they serve. This is a more responsible future, and a far less dangerous one.

Let me elaborate.

Spotlight

From Kiki's Life

As a kid, I was an artist, a writer, dancer, an actress, and a wannabe model (if you ask me nicely, I might show you one of my headshots!).

I went to college for computer science because I wanted to create special effects for the movies. By the time I really understood what it would take to get a computer science degree, it felt too late to back out.

I was a junior in college the first time I heard the term *unicorn* (and no, it has nothing to do with the high '80s ponytail in my modeling headshots). I was working at the local television station creating graphics for commercials and trying to get credit for work study. My professor, blown away by the fact that I was a "creative" who could also write code, paraded me around to others in the building, showing me off like a circus act. While I was obviously *allowed* to be in that program, I didn't feel *included*. Instead of belonging, I felt like an outsider who amazed everyone when I was anything better than mediocre.

Eventually, I transferred out of the department and made the decision to change majors, the way so many other women in computer science have done.

I'd love to say the reason why I left CS that first time had nothing to do with being female, but as one of only three girls in my core classes, I'm sure that's not true. Having so few women in my cohort meant I often felt the need to represent my entire gender. When I decided that I wanted to graduate early, it took much less energy to transfer to mathematics (where I was one of 10) to finish out my degree.

As fate would have it, this all contributed to my ability to return later in life, where I ran the Women in Computer Science club while earning a master's in CS at the University of Oregon (and I'm happy to say that I didn't have the same problem that time).

Would I have ever attempted computer science if I hadn't known it was connected to filmmaking? Hard to say. My dad had been involved in technology since I was very young, so it's possible that I would have recognized it as a valuable profession—but I can't believe that I ever would have gone at it with the same passion and drive if I hadn't thought that it connected so closely with my dreams.

Educator Takeaway: Computer science might feel far removed from the subjects that you're passionate about, but broadening student horizons just might help them see something in themselves that they wouldn't have otherwise known was there.

How Implicit Bias Can Work Against Our Best Intentions

Of all students taking foundational computer science in high school, roughly 68% identify as male. This points to a wider societal issue that could be keeping other members of our population away. As it turns out, several studies validate the idea that there is bias in schools around "natural talent," based largely on gender[5] but also on other demographic identifiers. This bias affects teachers and classmates and trickles down to individuals, causing many students to believe that they don't belong in the sciences—even before they understand what the classes entail. This infectious bias can actually lead students to believe they are "bad" at something they haven't even tried. These biases don't usually happen consciously; therefore, one term for the phenomenon is "**unconscious bias**."

Of all students taking computer science in high school, roughly 68% identify as male.

Unconscious bias (also called **implicit bias**) refers to subconscious attitudes and/or stereotypes that influence our thoughts and behaviors toward certain groups. Usually, we have no idea these biases exist, and most of us would be horrified to know how strong our biases can be—even those of us working in the equity space (and especially those of us within education). These biases are formed via personal experiences as well as exposure to societal messages and can have a profound impact on our decision-making processes. Research has shown that these biases can contribute to discrimination and disparities in various domains, including education, criminal justice, and employment.[6] Understanding and addressing implicit bias is crucial for promoting fairness and inclusivity in all aspects of society.

Many of us have a hard time recognizing our own implicit biases—or even acknowledging that we may be under the effects of powerful implicit biases, because our conscious minds are so well-intentioned. This can lead to two opposite, but equally troublesome outcomes. The first is that we as educators may unconsciously allow biases to lead us to provide white males with higher grades and additional enrichment opportunities. The second is overcorrecting and taking pains to publicly highlight and overencourage nontraditional students—in the process unintentionally drawing attention to the fact that the world expects them not to succeed (like in my "unicorn" story). Both outcomes have been shown to result in the belief that traditionally underrepresented students don't belong.[7]

It's in Your Hands Now

Teachers in K–8 are set up to have the most effect on a student's future, not just because most students determine what they're capable of by the time they reach the second grade,[8] but because schooling is generally a requirement at these levels, so 90% of students worldwide receive a primary education.[9]

Play Time

Most of us like to think that biases don't affect us—but we're wrong!

Give a gift to yourself and your students by looking at how your biases compare to the rest of the world's. Visit https://bit.ly/3FKuArz and choose the "Gender Science IAT" option to take their implicit bias quiz. When you're done, you'll see the up-to-date graph of how other users scored. Beware: Your results might surprise (or even upset) you.

When you're done, you can visit Chapter 7 to find methods of conducting normal classroom activities in ways that reduce unconscious bias.

Want to truly understand the harm that unconscious bias can do? Take a look at this article:

Four Ways Teachers Can Reduce Implicit Bias (berkeley.edu)

https://bit.ly/3SnelIQ

Secondary teachers also have a critical role, because this is the time when students gain the foundational skills that prepare them for higher education, and that preparation can help them achieve a sense of belonging. With *just a little extra work*, we have the opportunity to bolster computer science as an option for children in all communities.

As an educator, even I reread that last sentence: "Just a little extra work." What does that mean, exactly? Teachers are already overworked and underpaid, with nearly 24% of working hours happening at home.[10] This means that even a little extra work is loaded onto a heap that's already spilling over. That's why CSEd integration has become so popular over the last few years. If teachers can replace some of their current workload with lessons that achieve more in the same amount of time, then it should be a net gain.

Plus, the integration of computer science as a tool in familiar courses helps to introduce the subject as a form of creativity and self-expression. It stands to reason that teachers who successfully incorporate computer science into classes that students already love are most likely to develop students who are prepared to meet CS requirements before graduation.

Adulthood will look very different for our students than it does for us. Someday, self-driving cars will flood the streets, exploration of the universe will be more achievable (both physically and virtually), and robots will inhabit most households ... not just to share the weather, but to help out with vital chores or monitor safety issues. The general public needs to be equipped with the knowledge and ethics necessary to make sure those inventions help more than they harm.

As educators, we can work together to create exciting and inclusive opportunities for students to shape their own worlds through computer science integration.

I'm not saying that this transition is going to be effortless. There will be challenges, glitches, and moments when you feel like tossing your laptop out the window. But it is in those moments of frustration that true growth happens—and if you're willing to take those first steps, I'll do my best to prepare you for the journey ahead.

Summary

Rather than overshadowing core areas, computer science integration is meant to enrich the understanding of multiple subjects simultaneously. The goal of CS integration is to equip students with valuable skills that enhance problem-solving abilities and future readiness, while still maintaining the integrity of your original lesson plan.

When done well, integration empowers a wider range of learners to thrive in an increasingly digital world, fostering diversity and inclusivity in the process. Your hard work might lead to the discovery of someone else's passion!

Reflection Questions

1. Did you take the implicit bias test? If so, how did you feel about the results? Can you point to any variables in your life that may have skewed your test? If not, what do you think is holding you back from giving it a try?

   _____

   _____

   _____

   _____

2. Can you recall an event in which you had every right to participate where you felt like an outsider looking in? How did that affect your attitude toward the group or subject associated with that event?

   _____

   _____

   _____

   _____

3. What are you most concerned about when it comes to integrating computer science into your subject?

_____

_____

_____

_____

What Teachers Are "Supposed" to Know About Computer Science Before Integrating

CHAPTER #2

Q: Exactly how much are we expected to know about CS before handing out laptops and tossing our students into the computer science pool?

A: About 50 minutes' worth of material more than what your students know.

You do not need to be a fount of knowledge to teach computer science. In fact, you only need to be an idea sprinkler to get the process going and initiate learning. Computer science is less about pouring facts into a student's mind and more about modeling curious behavior. After all, much of what a student is taught about coding in the fifth grade might no longer be true by the time they graduate from college. For that reason, it's far more important that you teach K–12 students to figure things out on their own than it is to teach them how to code in any specific language.

> *It's far more important that you teach K–12 students to figure things out on their own than it is to teach them how to code in any specific language.*

It may feel strange to be only one CS lesson ahead of your students, but next year you'll be at least a term ahead of the incoming class, and the year after that you may be so comfortable that you're able to develop custom CS lesson plans. You only have to take that first step once. After that, you'll learn which questions to ask, and experience begins to build quickly.

With that caveat in place, here are the four basic principles that you'll need to be comfortable with before you can host a successful coding or computer science class within your home subject:

Principle 1: Attitude Is Everything
Principle 2: Be a Lead Learner
Principle 3: Diversity Matters
Principle 4: Understand Concepts

Let's dig into each one of those ideas.

Principle 1: Attitude Is Everything

Q: I am nervous about trying to fit computer science standards into my lessons. What if coding is too hard or boring for my class and the mood in my room suffers?

A: There are a lot of fun and authentic ways to bring computer science and coding into the classroom. You don't have to start out with tools that are repetitive or overly complicated. Take some time to look into what other teachers are finding success with and try adapting those for your lessons. There's absolutely nothing wrong with looking at CS activities as "fun-time" or a reward within non-CS classes. When I was in school, we played games with paper and pencil to reinforce what we were learning in the classroom. In the 21st century, you can play with micro:bit or Arcade. The joy that comes from coding starts with you!

It may feel strange to be only one CS lesson ahead of your students, but next year you'll be at least a term ahead of the incoming class, and the year after that you may be so comfortable that you're able to develop custom CS lesson plans.

You have probably already noticed that kids can be unusually intuitive, especially when around someone for long periods of time. You may find that some of your students are particularly in-tune to your moods and attitudes, whether they choose to use that awareness for good or for evil.

This is why your attitude is so important when introducing computer science and coding to your classroom. If you truly believe that you are presenting your students with a gift by helping them get comfortable with CS early in life, they will feel that. If you are happy to present them with opportunities to code because you see the tool as a useful and expressive form of art that empowers students, they will feel that too. Conversely, if you see CS as a burdensome subject that you've been forced to find time to fit into your lessons and you are eager to get it over with, you can be sure they'll pick up on that just as clearly.

Studies have found that mentorship is a notable factor for inspiring students to choose computer science, particularly with girls and nonbinary participants.[11] The power of having someone represent CS as a positive and achievable option creates a draw and allows students to see themselves participating in the craft, especially if their experiences happen prior to puberty while children are still forming beliefs of who they are and what core strengths they have.[12] For this reason, your best shot at getting students to believe that computer science has an important role to play in their future is for you to believe it first.

The importance of attitude is also present within each component of practicing CS. Whether it's dealing with frustration, disappointment, or **fiero**,[13] the way you approach your emotions as they relate to code (yours or theirs) will go a long way toward shaping what students believe is acceptable within themselves. In fact, coding is a fantastic—and authentic—way to introduce discussions on **emotional literacy**, **growth mindset**, and teamwork.

"Fear has a smell, as love does."
—Margaret Atwood

Computer science, like the other sciences, requires a great bit of trial and error, exploration, and persistence. For the importance of these techniques to land, students must feel safe failing fast and failing often. This can be contrary to what they experience in a class like mathematics, where students are often expected to have correct answers on first submission. Your attitude can support students in this journey, as well.

One of the most crucial attitudes to be aware of in relation to CS is your outlook on **frustration**. Frustration is a fundamental part of learning, and research shows there is a sort of continuum of intensity for frustration that drives a learner either to explore, ask for help, or give up entirely.[14] Frustration is a natural response to lack of understanding, and the process of gaining understanding alleviates that frustration and initiates learning. Intentionally pointing out that relationship can comfort a student and help them perceive the signs of frustration as precursors to success.

Oddly, frustration is also a fundamental part of having a good time![15] This might seem counterintuitive, but for humans to have long-lasting fun, they need to repeatedly experience the cycle of frustration and fiero. *Fiero* is an Italian word, believed to be coined by psychologist Isabella Poggi to indicate an intense pride felt after a hard-won achievement. It is said that one cannot experience fiero without first feeling frustration—the more profound the frustration, the more rewarding the fiero. This is similar to what one might experience when making 10 failed batches of meringue cookies before perfecting their first dozen.

Often, as teachers, we try to overprepare students for the assignments they're about to undertake to prevent frustration altogether. But until students *experience* frustration, they have no idea *why* they would want to know what we're trying to teach. This is my argument for encouraging you to educate students about frustration—including signs to watch out for, tools for processing, and empathy for others who are experiencing it—instead of trying to prevent it.

In my experience, when I have had deliberate conversations with my students connecting the experience of frustration to the outcome of success, students are less likely to internalize multiple failures as a reflection on their own intelligence and are more likely to view the entire process as a valid approach to learning.

This topic could almost be a book on its own. To learn more about dealing with frustration in education, take a look at my blog: https://medium.com/geek-groupies.

https://medium.com/geek-groupies

To read a QR code, you must have a smartphone or tablet with a camera. We recommend that you download a QR code reader app that is made specifically for your phone or tablet brand.

Principle 2: Be a Lead Learner

Q: I haven't had any official computer science training, so I'm worried it won't be long before my students know more about the subject than I do. How can I avoid looking silly when they ask me questions to which I don't have answers?

A: The same way you would if they asked you a tough question in your subject of mastery! Honestly, thanks to inquiry-based learning, it's nearly impossible for a student to tell what teachers do and don't know (unless the beads of sweat trickling down your forehead give you away!).

Oddly, frustration is also a fundamental part of having a good time!

In my days with Code.org, we intensely championed the idea of being a "Lead Learner." It wasn't just a title; it was a lifestyle!

Being a Lead Learner means you're embracing the idea that you are not the ultimate source of knowledge as it relates to computer science, but more importantly, that students don't *need* an ultimate source of knowledge to learn something new. In this way, you flip from a "teacher" to a "guide," getting comfortable with the phrase, "I don't know ... how can we find out?"

Personally, I call this "teaching like a grandma," but scholars call it **active learning** or **inquiry-based learning**. It's something that most teachers are comfortable with when it comes to helping students engage in the problem-solving process, but that instinct can fade when someone feels like they need to prove themselves. Somehow the less one knows about a subject, the more they feel like they should be able to provide detailed information in a moment's notice. That's neither helpful nor effective. Instead, encourage students to do their own investigating, and try taking the student empowerment even further by letting them know that you're learning alongside them, and you'd like them to teach you what they find out. You're not a Sage on the Stage; you're a Lead Learner.

Introducing the Lead Learner persona can help with the fear of being taken by surprise. If, for example, someone asks you the difference between a function and a method, you don't have to feel unprepared. You'd be doing them a disservice to tell them. They need to know how to discover the answer on their own, because someday that answer might change, and they will need to be able to figure that out. This inspires students to turn their brains back on and take responsibility for their own learning, while removing the unrealistic bar preventing a social studies teacher from bringing MakeCode Arcade into their unit on early Mesopotamia.

(The answer, by the way, is that there's practically no difference between functions and methods these days, depending on the coding language. Methods, functions, subroutines, procedures, etc., are all basically the same thing, though I'm sure there's a nitpicky professional programmer out there who will be happy to share some tiny nuances if you ever bring it up at a party.)

Spotlight

From Kiki's Life

In the early days after forming my educational nonprofit, Thinkersmith, I used to visit elementary classrooms and teach lessons on computer science for free (with the caveat that teachers stay and watch to replicate activities in future classes). I would often teach a series of eight to 10 sessions, and I never knew what kind of atmosphere I would get. Some groups were disciplined and routine-based, while others were more wild and unfocused. Each of these class types had pros and cons when it came to exploring the world of CS.

One day, I stumbled across the ultimate Lead Learner activity that changed the trajectory of my lessons forever.

During the fourth week of classes in 2012, I introduced a lesson called "Thinking Myself." This was based on a web game that I had created for my master's thesis, where the user had to figure out how to win a game that had no instructions and no clear indicators of what to do or where to start.

Through development and research, I discovered that the time it took to "win" this game was proportional to the age of the player. In other words, the older you were, the longer it took you to solve the puzzle. As it turns out, we develop unnecessary rules and biases as we age, making it harder to learn using real-time feedback. The older you are, the more likely you are to draw on previous knowledge, whereas younger people are more able to learn through exploration.[16] I wanted to promote this exploration for my CS students and teach them to embrace the journey, so I figured this game would be a worthwhile exercise.

I started this lesson with a chat about frustration (as I encouraged earlier in this chapter), giving students an example of things that are easy, which rarely bring joy—like brushing your teeth. I then compared those experiences to difficult things that are fraught with frustration and described how those lead to the development of excitement and pride of mastery. I encouraged them to think of frustration as a sign that they're about to learn, and they should celebrate themselves for that.

Next, I prepared them for the fact that I wouldn't be providing *any* help during the lesson. I told them that they were going to feel stuck, they were going to have questions, and they were going to say, "Ms. Kiki! Ms. Kiki! I need help."

And I told them that I wouldn't give them a single answer or instruction, but I wasn't being mean—I was showing them they have the power to learn something, *even when no one teaches it!* Then I released them to the activity.

Students spent about 15 minutes in pairs playing online with a machine that did nothing (see Figure 2.1), until the correct knob was turned—then a single green light would illuminate. Their next move might cause the light to

Figure 2.1   The Machine With No Instructions From "Thinking Myself"

Source: Kiki Prottsman 2011. Created using Thinking Myself.

turn off—unless they flipped the correct lever—then a bar on the side would start to climb, indicating progress. Every time they were wrong, the game started over.

Before long, students discovered that they needed to fiddle with each different element in the proper order until a meter on the side reached the top of the screen and they won the game!

After this exercise, we took a few minutes to discuss the thoughts and feelings that students had during the experience. We took stock of the moments of frustration and how great it felt to finally succeed. I would ask them to remember that emotional progression for future lessons, because frustration often leads to learning, which leads to success.

When I tell you that my students were never the same after that lesson, I am not exaggerating. Since that day, I have used a similar "Thinking Myself"–style of lesson with every classroom at every grade I have taught. Each time, the students have a shift in outlook that makes the rest of our sessions exponentially more effective. Students become more willing to explore, investigate, and solve their own problems, instead of expecting me to pour information directly into their heads. It's absolutely incredible.

Educator Takeaway: Frustration doesn't have to be avoided entirely. Students might end up surprising you with their ingenuity and persistence if you prepare them for the big feelings that they're about to have before you assign them difficult puzzles.

The trick to being an effective Lead Learner is making sure students know when you are wearing that persona so they're prepared to do additional work themselves. I used to wear a special "Lead Learner" T-shirt for those lessons, so students would recognize that my answers to their questions would likely also be in question form.

"Tell me and I forget, teach me and I may remember, involve me and I learn."
— Unknown (often attributed to Benjamin Franklin)

Source: Kiki Prottsman, 2024. Created using Midjourney and edited with graphic design programs.

Principle 3: Diversity Matters

Q: Shouldn't we let people make up their own minds about the need to study computer science? After all, if there are groups of historically underrepresented individuals who aren't interested in computer science, who are we to force their hand?

A: Making sure that all students have the opportunity to be introduced to high-quality, fun, meaningful computer science is not the same as forcing their hand. Coding is a hidden profession that many may not otherwise realize is an option until they've already convinced themselves that they're not technical. Computer science education is about even more than that. Careers in technology pay well, the number of job prospects are growing, and—most importantly—computer science teaches a set of skills that will be baked into nearly every other **STEM** (science, technology, engineering, and math) job likely to appear over the next decade.

You may be asking why I would continue to call out diversity and inclusion within a book about computer science in the K–12 classroom. Indeed, most schools represent the communities in which they are located, so it would stand to reason that each class within a school does the same. This is not the case. As of the time of this writing, computer science is treated as an elective in most schools, so it's up to students to self-select into classes based on course descriptions and feedback from peers.

Before we dive deeper, I want to be upfront about my intentions. This section is not a political statement, and I will refrain from including biased viewpoints or arguments based on my own personal beliefs. Instead, I want to look at this from an educational perspective, focusing on ideas, numbers, and facts as they relate to student experience and workforce impact.

According to *Business Insider* magazine, the #1 best job prospect between 2021 and 2031 is software programming based on median pay and industry growth.[17] This is not to say that we want all our students to graduate and immediately move into the tech industry, but the Bureau of Labor Statistics hypothesizes that STEM jobs will grow 10.8% by 2031, compared to 4.9% for non-STEM jobs,[18] and about half of all STEM jobs require understanding computer science.[19]

> *If we want to ensure the economic success of our students, we need to prepare them for the world that they will be adulting in, not just the world we are in now—and that future world will ooze with tech.*

Even for entrepreneurs, the ability to put together a simple website, app, or dynamic spreadsheet requires the comprehension of a nontrivial amount of computer science, and that means students will need to keep their skills sharp as technology changes around them.

If we want to ensure the economic success of our students, we need to prepare them for the world that they will be adulting in, not just the world we are in now—and that future world will ooze with tech.

With the knowledge that tomorrow's adults will need to understand hardware, coding, and data processing in ways that we've only begun to explore, it would be absolutely unfair to thoroughly enrich the lives of students who self-selected, while ignoring those who had not yet made the connection. That would be like only teaching students to write if they already wanted to be an author when they grew up.

All of this is deeply tied to equity, diversity, and inclusion. All students deserve to believe that they belong in the classes where they can learn these fundamental skills.

Before we go any deeper, let's define exactly what I mean when I use these common terms:

> ◗ *Equity:* Giving everyone what is needed for them to meet the desired objective. This could mean paying more attention to students who don't have help at home, allowing more time for students who have jobs outside of class, or adjusting the end-goal of students who started further behind than the rest. Equity differs from **equality** in that you are not treating everyone the same; you are giving more to students with

disadvantages so they can catch up to their peers. It's similar to starting runners closer to the finish line when they are racing in the outside lanes or having a score differential in golf.

> *"Smart teams will do amazing things, but truly diverse teams will do impossible things."*
> —Claudia Brind-Woody

- *Diversity:* Having a team made up of a variety of people from different cultures, genders, backgrounds, and with different perspectives. This nonhomogeneous mix helps to ensure that multiple viewpoints are considered, creating products that appeal to more people overall.

- *Inclusion:* More than allowing everyone a seat at the table, inclusion means that you are *welcoming* everyone to the table and providing a culture that allows participants to be their true selves without having to pretend that they have assimilated into a dominant ideal.

Equity and inclusion can be thought of as methods of increasing diversity, which is not only important for historically underrepresented individuals who want to join a space, but also for the groups that they belong to. A myriad of papers have been written on the way diverse groups come up with more creative ideas and more successful products, so long as the environment is considered inclusive and the teams feel they are part of a psychologically safe climate.[20]

In Chapter 7, we'll do a deeper dive into why diversity matters to the tech industry, as well as describe various ways to recruit and retain more students from historically underrepresented groups into tech classes.

Principle 4: Understand Concepts

Q: Do I need to know all the technical terminology used in computer science so students can begin to use this vocabulary to talk about what they're learning?

A: Do you need to use the word *algorithm* with kindergarteners to teach sequencing? No. Is there a benefit to introducing formal CS jargon in a class where students feel comfortable, so the words seem less intimidating later? Absolutely. The sad truth is, if you never call if/then statements **conditionals**, students will never realize that they've learned conditionals (even though they've been practicing them since the second grade) and they won't understand that they're much more prepared for a coding class than they think they are.

Similarly, *you* have likely already learned most of the fundamental concepts of computer science, though they haven't been connected to the corresponding terms in a computing environment, so those concepts might feel newer and more foreign than they truly are.

https://qrs.ly/9ofbmr7

The Lead Learner persona will help you get away with a lot, but if you're the type who likes to really understand your lessons before you teach them, there are some key terms to become familiar with before you get started. I attempted to list them all here, but it made for a long and slightly awkward chapter, so I've moved them to my blog: https://medium.com/geek-groupies. If you feel like doing some extra credit, try hopping over there to learn more about computational thinking, debugging, variables (the computer science version), functions (the computer science version), "Mean" (Taylor's version), and more.

I've added a whole bunch of other words into the glossary as well, in case you'd like to look anything up while devouring the rest of this book.

Summary

You don't need to be an expert in coding to bring computer science to the classroom; you only need to know how to foster a learning environment where teachers and students are co-investigators. Remember that attitude is important—your mindset can influence student engagement and learning outcomes. That said, frustration can be a precursor to success and emotional growth. Addressing frustration head-on before students encounter it can be helpful in teaching them how to persevere and push through until they learn what they need to know.

While computer science can open doors to well-paying STEM careers, that's not the only reason why all students deserve to learn CS. It's more important to prepare students for the world that they will live in when they grow up, whether they'll be creating tech or simply using it.

If you're eager to learn more about these topics and feel like a support system would be helpful in broadening your understanding of the subjects, continue into the next chapter to read more about developing comprehension and community in this space.

Reflection Questions

1. Take a moment of introspection to honestly explore your attitude toward bringing computer science to your classroom. What are four words that summarize your feelings around the possibility?

    _____

    _____

    _____

    _____

2. The shift from "Teacher" to "Lead Learner" can either be terrifying or liberating. What are some benefits that your students might realize due to being "guided" instead of "taught"?

_____

_____

_____

_____

3. _Equity, diversity,_ and _inclusion_ have become charged words over the last decade. Politics aside, can you come up with three beneficial community outcomes that may result from ensuring that CS students from historically underrepresented groups feel inspired and welcome within the computer science classroom?

_____

_____

_____

_____

4. Look through the fundamental CS terms explained in this chapter and identify the one that you understand the least. Work to find a different real-life example for the concept and write it out based on your understanding of the idea. (Feel free to use YouTube and **ChatGPT** to clarify any misunderstandings.)

_____

_____

_____

_____

CHAPTER #3

How We're Supposed to Know What We Know About Integrating Computer Science

Q: I feel like my colleagues just "get" this, and sometimes I struggle even to understand what the administrators are talking about when they say they want to promote algorithmic thinking. How can I quickly get on the same level as my peers?

A: Don't be fooled. A majority of teachers across the world don't feel comfortable teaching computer science or coding. That's one of the main reasons why the first K–12 CS movement was crushed in the late '90s and why it's taking so much time and resources to get it moving again. It's not for lack of interest, but lack of self-efficacy … which is only natural, since many of the educators of this generation didn't experience computer science classes themselves growing up.

When you think about it, we know what K–12 language arts looks like because we took K–12 language arts. We know what K–12 physical education looks like because we took K–12 physical education. Very few K–12 teachers took computer science while we were in school, but now it's up to us to determine what it will look like for the next generation.

The last chapter went over the basics of what you need to know before you start adding computer science to your coursework. In this chapter, I'll cover some resources for you to lean on when you get yourself into a sticky pickle—or, you know, if you'd like to preemptively deepen your computer science understanding to the point where you comprehend enough to be obnoxious.

"With great power comes great responsibility."
— Spider-man's Uncle Ben

Free Online Computer Science Activities

For adults, I'm not a huge proponent of jumping right into computer science at an age-appropriate level. Why not? Because I believe that early computer science is 90% confidence and 10% **grok**. (I believe that's also true for the early stages of most subjects, but this isn't a psychology book.) If we want to truly feel like we belong in CS—if we want it to spark joy and ignite a desire to keep us moving forward—we must believe that we're good at it, and it must be fun.

As I mentioned earlier, you don't need a tall stockpile of CS knowledge to bring edifying activities to your students, but if you find the subject personally enriching, or you are looking to move your courses more toward STEM (science, technology, engineering, and math) or **STEAM** (science, technology, engineering, *art*, and math), then you may want to use this chapter as more of a guidebook than a reference.

I could suggest jumping right into a free, online college computer science class (there are a few recommended at the end of this chapter), and I'd be confident that you'd come out the other side able to write a program—but I'm not sure that would help you *love* CS.

I've had significantly more success asking teachers to pretend they're kindergarteners and live the experience of learning to type with all 10 fingers before asking them to pretend they're second-graders playing a sequencing game online. Next, they're fourth-graders creating video games with blocks in MakeCode, then eighth-graders doing Intro to Programming on Khan Academy.

When I use this method, I start you at a level well below where you need to be. In this way, success comes naturally and you're able to safely fortify your knowledge base before you start stacking harder experiences on top. This approach also allows you to experience each lesson through the eyes of a fictional persona, so any insecurities or uncertainties can be offloaded to a faux character and you don't have to carry those along to the next activity. Think of it as educational regression therapy.

Now that you know my trick, I'm going to ask that you allow yourself to set aside any thoughts about where you *should* be, and instead, start at the very beginning so you can be confident that you have experienced every important landmark along the way.

> *I'm going to ask that you allow yourself to set aside any thoughts about where you should be, and instead, start at the very beginning so you can be confident that you have experienced every important landmark along the way.*

Feel free to get deep into character with these activities. You can choose one from each category, or do them all ... but whatever you do, try doing it through the eyes of a student who is experiencing the tasks for the first time. This can also help build empathy for your students as you begin introducing similar lessons into your classroom.

Note: I'm not necessarily advocating for you to assign any of the following activities to your students. These are less about "integration" and more about "preparation." Use these exercises to help yourself go back in time and create the computer science foundation that you might have had if you were growing up today.

Step #1: Kindergarten-ish

In practice, I prefer that kindergarteners focus mostly on **unplugged** lessons and the building of persistence and problem-solving skills. That said, to prepare for life down the line, this is also a great place to start introducing small quantities of keyboarding and logic.

This level is all about using the computer and strengthening the ability to recognize patterns while practicing fine motor skills. Since technology sites come and go, I'll provide some key search terms in case you need to find new games at this level: *free typing games, drag and drop practice for kids, free coding for nonreaders.*

1. Home Row Typing with Squirrel: https://bit.ly/3QdhXdy

    An absolutely adorable game where you practice typing single letters with the help of a hungry squirrel

2. Dragon Drop: https://bit.ly/3FypgHJ

    Click on a friendly dragon and drag him around obstacles to find his dinner

3. CodeSpark Academy: The Foos: https://codespark.com/play

    Use picture tiles to code your Foo and get them to the donuts (You will need to create a free account and scroll down to "Donut Detective")

Source: CodeSpark.

Other than the obvious (keyboarding/mouse-use/coding), what skills are little ones practicing when they play games like these? Write a few words about skills you think transfer from these activities into other core elementary classes.

_____

_____

_____

Step #2: Second Grade-ish

This level is about introducing logic, problem-solving, and persistence. If the following games aren't accessible, here are some search terms to find new options: *free logic building games for kids, free reasoning games for kids.*

1. Water Sort: https://bit.ly/45V2STV

   Pour water of the same color into tubes

2. Asphodel Follows Directions: http://aka.ms/asphodel

   Carefully place arrows to get Asphodel to the trophy

3. Green: Paint Everything Green: https://bit.ly/45Ps4el

   Use intuition, trial and error, and logic to figure out how to paint everything green (this is one of those games that might be harder for adults than for kids)

4. Sugar, Sugar: https://bit.ly/3QBhUty

   Draw structures to get enough sugar in each cup

Play Time

From a Second-Grader's Perspective

Try playing "Green" from the perspective of a second-grader. What skills will they need to have before they tackle it? Can you imagine how much preparation is too much? Can you imagine how quickly solutions will spread once a single child figures out how to pass each level?

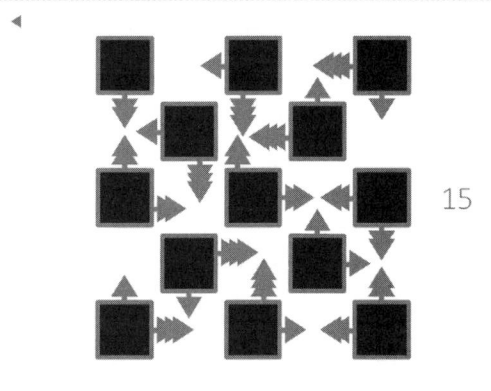

Source: Bart Bonte.

Do you think students will feel more proud of themselves if a peer shows them how to get past a difficult puzzle or if an adult helps them out?

Can you see how games like these allow you to expand your own frame of reference by starting with ideas that you are already familiar with before challenging you to test the limits of what comes naturally? How do you think you can use playful experimentation to create more opportunity for **constructionism** in your classroom?

_____

_____

_____

Step #3: Fourth Grade-ish

This level is about becoming familiar with sequence and structure. If the following games are not accessible, here is a term to search to find new options: *free **block-based** coding games for kids.*

1. Turkey Day: https://aka.ms/turkeyDay

   Code a turkey to rescue its friends and escape from the factory

2. Code a T-Rex: https://bit.ly/3tYtc2g

   Use blocks to code a T-Rex to gather all the coins in its path

3. NASA's Space Jam: https://s.si.edu/3MpiWq6

   Code your own solar system using blocks

Not all block-based coding opportunities are of equal quality. What's important here is that learners have guidance *and* creativity as they start to spread their wings and test the boundaries of what's possible. This is right around the age when we want to start allowing students to get themselves into trouble with their projects, because some of the best learning happens when trying to get back on track.

Source: Microsoft MakeCode.

When playing with the games above, did you experiment at all, or did you only do what the instructions told you to do? Why do you think that is? What can you do in your classroom to promote exploration and off-the-beaten-path thinking?

_____

_____

_____

Step #4: Eighth Grade-ish

Late middle-school is a good time to start taking coding and computer science more seriously. Students are generally better readers, they've had some foundational mathematics classes, and they're developmentally more capable of concrete thinking as they start to understand ideas that are more abstract. If the following activities are not accessible, here are some terms to search for to find new options: *free **text-based** coding games for kids*, and *internet safety games for students*.

1. Interland: https://bit.ly/45OrehR

   Play through minigames to learn more about internet safety and cybersecurity

2. Toxicode: https://compute-it.toxicode.fr/

   Learn to read code by following instructions with your arrow keys

3. CodeCombat: https://codecombat.com/play/goblins-hoc

   Choose your language and start traversing the map for loot

4. Chase the Pizza: https://bit.ly/3Sl6bkb

   Code a pizza-chasing game that you can share with friends

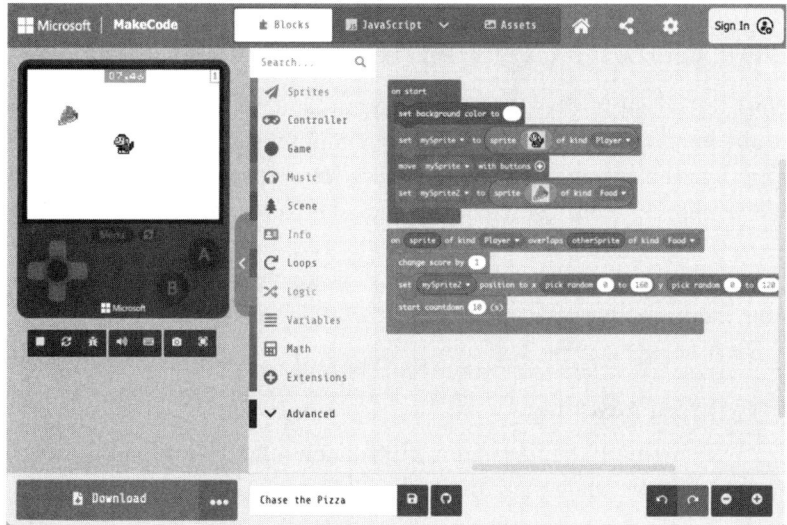

Source: Microsoft MakeCode.

To a student, text-based coding will feel like a completely different experience than block-based coding. Write a few sentences below describing a handful of ways that the two types of activities are similar.

_____

_____

_____

Step #5: Tenth Grade-ish

Remember that computer science is not just about coding. Students will need exposure to data representation, hardware, and networking, too. If these activities become no longer available, try searching for: _how a computer works_ and _data representation games_.

1. Inside Your Computer: https://youtu.be/AkFi9olZmXA

    A video that shows how the average computer works

2. **Binary** Numbers Game: https://bit.ly/46PW9vG

    Quickly calculate numbers in base-2 like your computer does!

We've barely scratched the surface of all that computer science has to offer, and it's likely you have some questions. Is there anything you're still wondering? Are there any fresh questions that have come to your mind since taking these steps?

_____

_____

_____

Free Online Computer Science Classes

Before you go any further, you should know that if you've come this far, you are already more prepared to integrate computer science into your classroom than the vast majority of educators. You absolutely _do not_ need to train any harder to bring life-changing lessons to your students.

If, however, you went through all of the exercises above and are feeling fairly confident about your ability to learn **foundational computer science**, there are several free and high-quality classes online.

Here are some of my favorites:

▶ **Khan Academy**

   Don't miss Khan Academy's introduction to the internet. It will take your understanding of digital information to the next level: https://bit.ly/479utSj

Interested in a survey of languages to help you program for the web? Try Khan Academy's programming class:

https://bit.ly/49dO1GY

▶ **Codecademy**

Want to dive into Python? Codecademy will walk you through the basics with easy-to-follow lessons:

https://bit.ly/40kIxWE

▶ **Free Code Camp**

With dozens of classes covering everything from web design to security to **data structures**, these courses can move a little quickly, so make sure to take advantage of their YouTube videos:

https://www.freecodecamp.org

▶ **Harvard**

Want to say that you learned to code at Harvard? Start by taking their free game development class, then move into one of their 15 other offerings:

https://bit.ly/40gibW7

▶ **MIT**

A little less intuitive to navigate, but equally promising, the MIT courses cover subjects like Computational Thinking and Computer Engineering:

https://ocw.mit.edu

Free Computer Science Educator Trainings

Are you finished bathing in the infinite firehose of knowledge around computer science? Come splash around beneath the CSEd spigot. There are slightly fewer resources out there when it comes to teaching computer science in K–12, but the information is thorough.

▶ **Code.org**

Code.org is the OG when it comes to online computer science professional learning courses. Not only are their modules free, but they contain several tidbits and takeaways that can be applied to any curriculum, not just the lessons specified on their page:

https://code.org/educate/professional-development-online

▶ **Microsoft MakeCode**

MakeCode has several different learning options, depending on whether you plan to teach with Arcade, micro:bit, or Minecraft EDU. I've put them all together into one handy landing page at aka.ms/makecode-pl.

*"If I have seen further,
it is by standing on the
shoulders of giants."*
— *Isaac Newton*

▶ **Pathfinders Institute**

Each summer, Infosys brings together some of
the best CSEd tools in the industry to provide
free online training to public school educators.
If you're interested in signing up, check for
open enrollment:
https://bit.ly/3FE8zuF

▶ **Ellipsis Education**

If you'd like to learn a little more at your leisure, visit Ellipsis
Education to watch age-differentiated webinars. They are
advertised as free, but sign-up is required:
https://bit.ly/3VTte7g

Computer Science and Tech Conferences

One of the best ways to truly absorb useful techniques and best practices
from veteran CS educators is to surround yourself with them in a collection
of tiny rooms for the weekend. That's where technical education conferences come in.

▶ **CSTA**

The annual Computer Science Teachers Association conference
alternates between even years in-person and odd years
online. Here, educators from across the United States share
ideas and enhance their teaching skills through workshops,
presentations, and insightful panel discussions. Truly one of the
best experiences for CS educators, CSTA makes everyone feel
welcome and appreciated. This conference puts a great amount of
emphasis on diversity, inclusion, and equity in CS, while striving
to make sure everyone has illuminating messages to take away.

▶ **ISTE**

The massive conference from the International Society for
Technology in Education, referred to as ISTE Live, is an
educator's paradise, drawing an international audience upwards
of 10,000 people each year. Stacked with goodies, their expo hall
is giant and includes several live demonstrations. For those who
have full-access tickets, there is a nonstop agenda of all things
EdTech. There is barely a CSEd tool or language in existence
that won't be represented at ISTE Live.

▶ **Bett**

Formerly BETT (British Education Training and Technology),
this organization hosts a series of global conferences that happen
across the world a handful of times each year. With over 30,000
attendees, Bett showcases the best tools and resources that

education has to offer. With a goal of improving outcomes for both teachers and learners, Bett strives to connect and empower teachers to provide better technological options in the classroom.

▶ **ITICSE**

The Innovation and Technology in Computer Science Education conference is a joint venture with the Association for Computing Machinery that claims to be the largest computing education conference in Europe. Recent programs show several sessions for higher education, with a handful of offerings for primary and secondary education.

▶ **ITEEA**

The International Technology and Engineering Educators Association conference is a global STEM event geared at training educators to bring computer science and STEM into the classroom. With a focus on bringing educators together, this conference has a little something for everyone.

There are also several smaller, local conferences across the globe, so if you're wanting to participate in something a little more intimate, try a browser search for "computer science education conferences near me."

Support Systems for Teachers Integrating Coding Into the Classroom

This may not sink in right away, but if you plan to integrate computer science into your classroom, then *you are a computer science teacher*. It doesn't matter if you only do a few lessons a year—if you are working to bring CS opportunities to your students, then you are bound to have many of the same doubts, journeys, and discoveries that any other CS educator would have. For that reason, I highly recommend that you embrace the title and recognize that there are groups out there meant to help you with your adventure. If you are feeling any type of **impostor syndrome** at this point, I assure you, there's no need for it.

While there are far fewer communities and resources targeted specifically at teachers practicing CS integration, many CSEd networks and activities developed specifically for K–8 will fill that need. When in doubt, target connections and exercises that focus on early CS education, rather than collegiate groups and activities. Here are a few organizations that welcome teachers no matter where they are in their CSEd journeys.

▶ **Computer Science Teachers Association**

Perhaps one of the friendliest and most well-known groups in the North American CSEd space, the CSTA has chapters all over America and throughout Canada, which allows them to nurture teachers locally as well as at the national level.

Spotlight

From Kiki's Life

I have a master's degree in computer science. I have eight (now nine) published computer science books. I developed the CS curriculum set that's used by the majority of elementary school children around the world ... keynoted conferences ... won awards—and I still often show up to conferences feeling like I don't belong. That's called impostor syndrome, and it tends to happen most frequently to women, particularly those in technology and education.[21,22] Impostor syndrome manifests as feelings of inadequacy, self-doubt, and the belief that personal success is merely a result of luck or deception.

Identifying impostor syndrome is the first step toward combating it, and combating impostor syndrome will be crucial for a more diverse landscape in the computer science field. Seeking support from peers, mentors, or counselors has been shown to be helpful,[23] and individuals can also challenge negative thoughts by focusing on their achievements and acknowledging that everyone experiences self-doubt from time to time.

If nothing else, I hope you come away from this book realizing that it's extremely common for those participating in computer science to doubt their own efficacy in relation to others in the field.

Educator Takeaway: If you're working to bring computer science into your classroom and you start to question whether you're doing enough or doing it right, try to focus less on the way your knowledge compares to the knowledge of others, and instead focus on how your understanding is increasing from day to day.

Source: Kiki Prottsman, 2024. Created using Midjourney and edited with graphic design programs.

The CSTA hosts one large yearly conference aimed at promoting computer science education in the classroom, but its many chapters also tend to host conferences and get-togethers throughout the year, allowing participants to really feel like part of something bigger. Their free membership includes access to their standards, local chapters, the ability to apply for conference scholarships, and a subscription to their newsletter that provides information on additional opportunities and resources.

For $50 per year (at the time of writing), teachers can sign up for CSTA Plus, which includes access to the CSTA learning series, access to recordings of previous conference sessions, access to professional development opportunities and courses, discounts on classroom resources, and more.

▶ **Canada Learning Code**

For teachers, Canada Learning Code is somewhere between a professional network and a set of professional development opportunities. With chapters across Canada, they aim to help teachers get comfortable bringing coding into the classroom. They have local meet-ups, host a virtual "Teachercon" each year, and have dozens of pay-what-you-can options for online teacher-facing trainings.

With lesson plans and learning tools, Canada Learning Code is a great, no-cost way to get your bearings as a CS educator.

▶ **Computing at School**

The UK's answer to CSTA, Computing at School has over 330 chapters (CAS Communities) across the United Kingdom. With virtual and in-person meetings, blogs, newsletters, educator toolkits, and more, Computing at School provides a vast network of peers to learn alongside and learn from.

The CAS website provides a bunch of no-cost lessons for all grade levels that can be downloaded under the Creative Commons Attribution-Share Alike 3.0 License, which allows for remixing and redistribution with proper credit.

▶ **International Society for Technology in Education**

While their name indicates global membership, ISTE's conferences and policies are generally focused on North America. In contrast to the other organizations mentioned, ISTE doesn't have local chapters, but they do have an "affiliate network," which is a directory of nonprofits that share in ISTE's mission to inspire educators worldwide to use technology and innovate in the areas of teaching and learning. For $85 (at the time of writing), they also allow basic members to utilize "ISTE Connect," which is a series of webinars and discussions geared toward education professionals.

You'll notice that ISTE's directive is generically more "tech" than "computer science" focused, but they have a huge number of opportunities online and at their yearly ISTE Live conference, where K–12 educators can learn more about bringing CS into the classroom. They are also extremely communicative via email with a slew of blogs, articles, podcasts, and webinars to keep educators up-to-date with all of the latest CSEd trends.

Summary

CSEd is loaded with support systems. Between online tutorials, professional learning opportunities, conferences, and consortiums, there are plenty of free resources for teachers who want to learn more about the subject. Don't be shy: Reach out!

Reflection Questions

1. This chapter was full of resources that can enrich your level of computer science knowledge. How much of it do you think you need to review to be comfortable bringing computer science to your students that very first year? Are you ready already? If not, do you think you would feel any more ready after getting a few certifications?

   _____

   _____

   _____

   _____

2. If anything is holding you back, what do you think that might be?

   _____

   _____

   _____

   _____

3. Did you review the age-appropriate lessons through the eyes of young students? If so, did you imagine any moments of misunderstanding or frustration? Capture those here:

_____

_____

_____

_____

4. Did any of the CS education organizations catch your eye? Try looking up professional networks around you and see if they have any free resources or nearby conferences that you can attend.

_____

_____

_____

_____

PART 2

A DEEP DIVE ON INTEGRATING COMPUTER SCIENCE INTO THE CLASSROOM

Let's mix it up.

"You make different colors by combining those colors that already exist."

—Herbie Hancock

CHAPTER #4

What You Need to Know About Integrating Computer Science

Q: I don't plan on teaching computer science, but I do want to add some CS to the class I normally teach. What should I keep in mind?

A: There's a freedom that comes with integrating computer science into other classes. When you're not trying to cover all of CS from A to Z, you have the ability to push forward and pull back in response to the comfort level of your class. That said, there are some best practices that work in CS that may collide with practices used in other classrooms. Ideals like: freedom to collaborate as needed, allowing learners to take multiple attempts at incorrect answers, and starting small may provide the smoothest experience for you and your students.

The last chapter offered a miniprofessional learning opportunity for educators who wanted to strengthen their own CS foundations before diving into integration. I like to think that the print experience contains many of the pieces that I would have covered in person, but with fewer jokes.

In this chapter, we'll shift the training more toward the *integration* piece.

Instructor Best Practices for Integrating Computer Science Into Other Classes

Teachers of core subjects get to introduce computer science in a more natural, less threatening way due to the fact that they can treat it like a novel addition to topics that students have been exposed to for years. But that's not to say that there's no risk involved. Those first experiences with computer science are likely to factor into a student's lifelong opinion on CS; therefore, it's important to follow best practices, even when integrating computer science into other classes.

> *"People sometimes talk about the power of first impressions, and believe me, there is truth to it."*
> — *Ann Brashares*

In my more than a decade of teaching and researching innovative computer science education, a few best practices stand out when it comes to incorporating CS into other lessons. These practices also tend to be best for CSEd in general, but I will particularly highlight those that feel relevant to integration.

Collaboration

Too many people think of computer science as something you do alone in a dark basement while eating Cheetos and drinking Mountain Dew. While I have known plenty of coders who could down a 2-liter of Dew while racing toward a deadline, I've yet to meet a software developer who works alone.

Computer science is a collaborative industry. Whether you're working in software, hardware, robotics, or the metaverse, communication and teamwork are vital components when it comes to the creation of successful products. Unfortunately, a lot of classroom work and assessment happen individually. In theory, this should give us a better idea of what each individual is capable of—but quality doesn't happen in a vacuum. It's about climbing atop ideas from others and having your own thoughts triggered by something that someone else says or does. It's about sharing your own process aloud so you can find mistakes and misconceptions as you hear the words coming out of your mouth. It's about testing things from your perspective and from the perspective of others.

Even if you aren't in a position to promote group work, consider allowing students to bounce ideas off of one another and ask each other for help during projects. Even better ... encourage it!

Productive Struggle

If you remember the discussion on frustration, fun, and fiero in Chapter 2, then you're already familiar with the idea that robbing a student of **productive struggle** is akin to robbing them of pride and enjoyment. But notice that I use the term *productive* struggle. There's a huge difference between the state of mind of a student who is actively flipping through pages and a student who has their head down on the desk in surrender. The trick is in finding ways to keep the struggle productive for as many students as possible for as long as possible. In my classrooms, I would weave a few different techniques together to make this feasible—a couple of which are centered around collaboration.

My favorite was spin on "Ask Three Before Me," and I call it, "Ask Thee, Ask Three, Then Ask Me." The title is long, but the results are pure gold. The idea is that a student should ask themselves aloud the question they need answered, and if they aren't satisfied with the result, they should ask two more classmates before flagging down the teacher. In my version,

I challenge a student to look at the problem in three different ways before reaching out to classmates with the quandary. When they do, they're prepared to help others see the problem from multiple angles as well, and therefore are more likely to come up with a solution before they ever need my help.

Another trick I use to promote productive struggle is "Red, Yellow, Green." In this classroom protocol, students have access to three colors of sticky notes (or cups, magnets, or other indicator). Periodically, you remind your students to update their status and each group will choose a color to indicate how they're feeling about the activity they're working on:

▶ Green—I'm feeling great, and I could help others

▶ Yellow—I'm doing okay, but I'm not sure if I could help others

▶ Red—I'm not feeling confident, and I could use help

By scanning these indicators, not only do you get an idea for where the class sits in their understanding, but other students can see who is displaying a green signal and reach out to them for help.

Quick accessibility note: Consider also having students write the words "RED," "YELLOW," and "GREEN" on their indicators so that those with colorblindness have an easier time choosing the correct flag.

Meet Students Where They Are
(AKA: Mind the Gap)

This practice incorporates a little bit of what you saw in Chapter 3, where I suggested that you regress to your kindergarten self before walking through the basics. When someone realizes that they have a gap in knowledge, they don't have any way of knowing how wide that gap is. They could be missing a single lesson or 3 years of intense study … but if the gap is perceptible, it can feel too wide to jump. The quickest way to help students become confident is to close that gap.

One of the most popular ways of doing this in CSEd is through the use of unplugged lessons. Most unplugged lessons incorporate some sort of game, craft, or other familiar activity as a way of drawing parallels between what students already know how to do and what they are going to be asked to do. These also make great equalizers between novice and advanced students, since people don't generally encounter unplugged lessons when they're tinkering around at home.

Another way to meet students where they are is to use what I call the **Escalator Method**. Just like an escalator is a set of interlocking steps that effortlessly moves you from where you started to where you need to go, the Escalator Method gradually moves students forward in such a subtle way that sometimes they don't realize how far they've come.

Spotlight

From Kiki's Life

When teaching, I like to practice something that I've dubbed "The Escalator Method."

Suppose you want to teach your class how to use **loops** in a computer program, but they haven't yet studied any CS concepts and they haven't done any programming. It would not be ideal to sit them at a computer with a set of instructions and expect them to pick up on the steps they need to follow and the reasoning behind using loops immediately. (I would call that the "teleportation method" because it would be ill-advised to make a jump that large since participants are bound to arrive on the other side with some vital pieces missing.)

Instead, you might first do an unplugged lesson (like "Circle a Chair" from Microsoft MakeCode). Or discuss loops in terms of the subject that you normally teach (relevant to music, physical education, art, science, etc.). This familiarizes students with loops as a concept, so they'll understand what loops are used for in practice, reducing their **cognitive load** during future activities.

Next, you might show students a program that "you wrote" using loops. Show them where the loops are, how you knew where to find the appropriate blocks or text to construct the loops, and how they could modify the loops to make them longer or shorter. This brings students one step closer, helping them to understand the layout of the computer interface that they'll be using *and* helping them transition the idea of loops into the context of programming.

Finally, you can show students how to get set up with their programming activity and allow them to follow the instructions as intended. Wrap up the entire experience with a review to put words to all the ideals that they've been working to commit to memory.

Educator Takeaway: You can teach a child almost anything if you start with concepts they already understand and bring them along on a step-by-step journey.

Source: Kiki Prottsman, 2024. Created using Midjourney and edited with graphic design programs.

Don't Be a Sage on the Stage

Computer science tends to bring out this weird behavior in some educators where they feel like they have to stand in front of the class and lecture while delving deep into picayune details to illustrate how much they know about the subject. This isn't fun for anyone!

Don't worry about trying to exhaust all the details about CS with your students. It's much more important that your students get to play and discover and have a good time with the material than it is that they come out of an integrated project knowing all of the various forms a variable can take or how many bits a **char** takes up in memory. This is the beauty of integrating CS into your classroom instead of teaching an advanced CS class. Embrace adventure, inquiry, and chaos over rote memorization.

Understand the Integration Landscape of Your School

Remember that whole thing about creativity not happening in a vacuum? That goes for educators, too! This entire experience can be made more meaningful by understanding CSEd that's happening in other classrooms. If you know that a neighboring science class plans to use a certain CS tool, see if you can incorporate that tool into your lessons, as well. If you learn that the language arts teacher plans to use MakeCode Arcade for literary projects, consider teaching students to use Arcade in your classroom first, so things go more smoothly at the end of their unit on Shakespeare. Not only will this help keep computer science from feeling siloed, it will also help students form a clearer understanding of the importance of CS as a tool.

If you're interested in learning more about CS best practices, there are a few resources that include compiled lists of the most effective methods as determined by research and/or classroom teachers. Two of my favorites (which have also captured versions of my tried-and-true best practices) are *Twelve Principles of Computing Pedagogy,* from the Raspberry Pi Foundation,[24] and *7 Research-Based Classroom Strategies for Teaching Computer Science,* from Engineering in Elementary.[25]

The Most Popular Tools for CS Integration

Before you start making decisions, it might be helpful learn what apps and hardware teachers love to use when integrating computer science into their classes. There are, of course, hundreds of possibilities—but I want to give you a taste of some of the most popular options so you can investigate these tools when preparing your own lesson plans and exercises.

Remember, when it comes to CS integration, it's not all or nothing! Integration generally allows you to try several options in pursuit of achieving required standards. This is great news when it comes to your first year, because it means you can take lessons one at a time and see which activities hit and which ones miss.

Much of what you'll find in this à la carte world will relate directly to coding, but every now and again, you'll discover a gem focused on cybersecurity, networking, or data representation.

Hour of Code

Where one-off experiences are concerned, it's hard to beat HourofCode .com. If you click the "Activities" link at the top, you'll get a giant list of lessons expected to take an hour or less. These activities can be sorted by grade, experience level, classroom subject, and more.

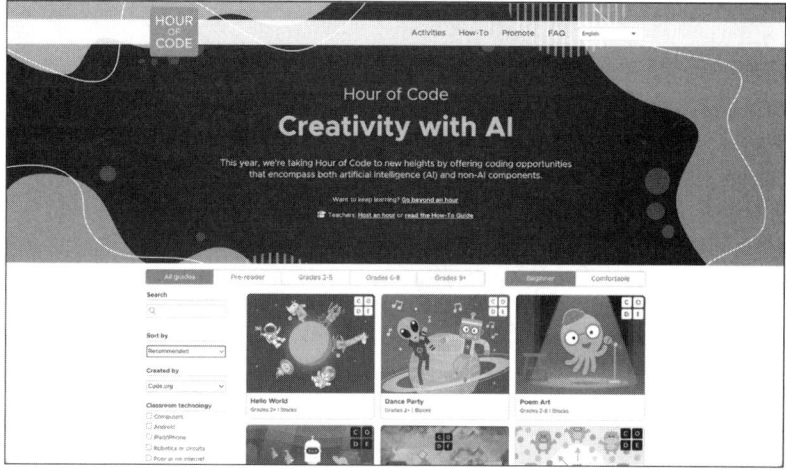

Source: Code.org.

When you first click on an activity tile, look for the "Teacher notes" link. This should lead to some sort of lesson plan and/or set of standards for the exercise. The Hour of Code site does not yet have a way to search by standard, so if you're looking to satisfy something specific, you'll need to use the traditional search field in the upper left corner of the activity area.

Play Time

Integrating Games

Hop over to https://hourofcode.com/learn. Use the filters to narrow down activities to an appropriate grade and subject for your classroom. Choose any of the top 20 games and give one a try. Can you conceive of a way that you might be able to blend the activity seamlessly into the lessons you already teach?

MakeCode

Of course, I also need to mention MakeCode! I'm not adding this to the top of my list solely because I currently work here. This is legitimately one of my favorite products when it comes to flexibility and pathways. In fact, the reason I ended up with this team is that I was digging into their offerings on my own when writing books and curriculum sets, and I got to know them really well. This product is still growing, and I love how they listen to the people who share on their forum (forum.makecode.com).

With multiple platforms (like Arcade, micro:bit, and Minecraft) plus multiple interfaces (blocks, JavaScript, and Python) it's hard not to love playing with the provided tutorials. They accept sign-ins via Microsoft, Google, and Clever—even though most activities can be done without signing in at all.

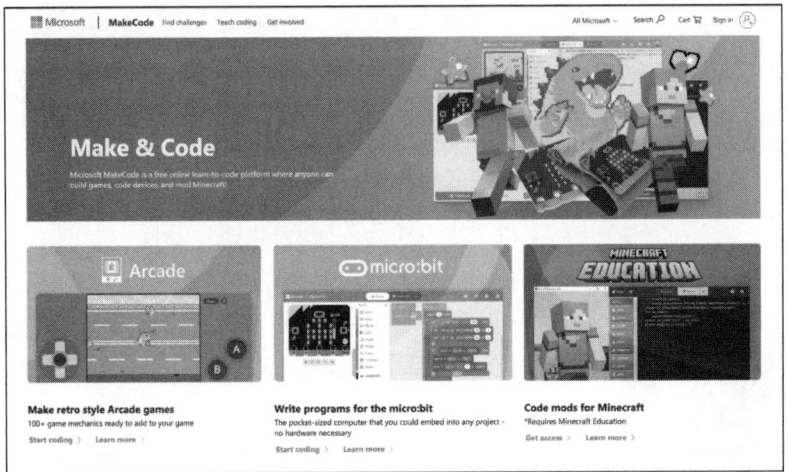

Source: Microsoft MakeCode.

Junior Block-Based Favorites

There are plenty of coding opportunities for little ones. While MakeCode is only just coming out with options for students in K–2, the following tools have been rocking the space for quite a while:

- CodeSpark: The Foos
- ScratchJr
- Kodable

Other Block-Based Favorites

We can't forget the classics! These tools have been staples in the process of learning computer science for a long time. Whether students are beginning in the fourth grade or 12th, block-based interfaces offer a solid start to any integrated activities:

- Scratch
- Alice

- Tynker
- Quorum
- Google App Inventor

Text-Based Favorites

If you find that your students already have a grasp of the basics, you can give them some text-based programs to play with. These options make it fun to learn text-based languages. Most of them are flexible enough to allow students to figure out how to create their own projects, but all of them make fantastic stepping-stools to the next level:

- Codecademy
- Pencil Code
- Khan Academy
- CodeCombat
- VS Code for Education

Classic Text Languages

You may find that you want to walk students into learning classic languages directly. These are the most popular languages of the day. (*Pssst* ... don't confuse Java and JavaScript! The names are similar, but they're really not related at all.)

Each of these languages has strengths and weaknesses, and when you're ready to bring them into the classroom, there's a *ton* of great information online around which are most popular in the classroom and why:

- C++
- C#
- Python
- Java
- JavaScript
- Swift

Robot Favorites

Companies are making robots now that can be used all the way down to kindergarten! Try bringing some robotics into the classroom for a fun **constructionist** project!

- Bee-Bots
- Birdbrain: Finch

- Sphero

- Wonder Workshop: Dash

- Ozobot

Hardware Favorites

Some of my absolute favorite activities include electronic engineering. And it doesn't have to be as hard as you might think! There are so many resources online to help you get started with any of these tools:

- BBC micro:bit

- Circuit Playground

- Makey Makey

- Chibitronics

- Raspberry Pi

- Arduino

Unplugged Lessons

And don't underestimate the power of introducing computer science without any technology at all!

- CS Unplugged

- Code.org Unplugged

If the trial-and-error discovery method isn't for you, you can also hunt down premade lessons by searching **Teachers Pay Teachers** using keywords from the standards you're trying to hit paired with the name of the class you generally teach. This search method also occasionally works on a web search engine.

Looking Into Curriculum Sets

Once you've moved past individual exercises (which could be quite a while if you are integrating activities) you may be at the point where you're ready to find a substantial set of lessons, or even a complete curriculum.

If you teach math or science, then you're probably going to have a slightly easier time finding a turnkey solution than educators in the humanities, but if you look around, quality resources are popping up everywhere. For example:

- The state of Virginia has required CS integration at the K–12 level since 2017, so it makes sense that they'd have a repository of cross-curricular, peer-reviewed lessons ready for download. You can find that library at GoOpenVA.org.

- Code.org, known for self-paced CS for all ages, hosts a "Connections" area that offers cross-curricular lessons in bundles at https://code.org/educate/csc. Here you'll find tips and tricks, as well as lessons that integrate with math, science, ELA, and social studies.

- The STEM Materials Center (SMC) at Educational Service District 112 in the state of Washington has also compiled a series of units meant to be integrated into core classes for a variety of ages: https://bit.ly/3tHtNVF.

- CodeHS (https://codehs.com/curriculum/catalog), Bootstrap (https://www.bootstrapworld.org), and ProjectGUTS (https://teacherswithguts.org) all offer well-respected, free resources for core classes.

- Barefoot Computing, an offshoot of Computing at School, has a nice subset of integrated lessons on their site at https://www.barefootcomputing.org. Their lessons can be downloaded with a free account.

Aside from what I've mentioned here, there are other little pockets of lessons that pop up randomly in searches. Some are good, and some … not so much. To help you tell the difference, I'm including a set of bullet points that you can use for evaluation purposes:

- Was it created by anyone mentioned in this book? If so, chances are that it's pretty good!

- Is it associated with an organization or university? If so, there's probably research involved. Go for it!

- Does the site provide a star rating or number of downloads? If so, look for lessons at 4-stars and above with at least 100 downloads. If 100 people have looked at the curriculum and there are no major complaints, it will probably do just fine.

- Does it have a table of contents or unit list? If so, that's a good clue that it will be well organized.

- Does it clearly list any standards? If not, it may take a while to process those yourself.

- Is it free? If not, make sure you can view sample pages to get an idea of whether or not it's high-quality.

If all else fails, put a call out to your network. If you belong to teacher groups on Facebook, Twitter, or Pinterest—if you joined a CSEd community like CSTA or Computing at School—there's bound to be someone who has experienced the same journey that you're looking to trek. Connect. Ask questions. Trade lessons. You don't need to do this alone.

Tips for Creating Your Own Curriculum

Computer science integration is not yet the global norm, and since different states and different countries hold different requirements, it can be hard to find lessons that will fit perfectly alongside your established activities. This section will provide you with some general tips for creating or adapting lesson plans for your needs.

Work Backward

You may already be familiar with **backward design**.[26] This technique is one of the best ways to ensure that your lessons meet the needs of your administrators. It also ends up being one of the easiest ways to load your activities with creativity and fun, because you get the "must dos" out of the way early and end up with an infinite number of ways to get there.

When using backward design, you'll want to start by getting crystal-clear on the standards and objectives that you want met. By identifying these in the beginning, you'll be sure that your lessons hit those marks when they're finished.

Next, figure out how you'll know that you've met those standards and goals. Will it be via student assessment? Will you simply need to observe that your students can accomplish something that they couldn't before? This bit may change during the creation of the lesson, and it may need to be updated after the lesson is tested, but knowing what you want to see will help you predict when the lesson is done.

Finally, you'll need to decide how to prepare your students to accomplish the task you laid out. This could be anything from memorizing a song to constructing an image out of puzzle pieces to creating an arcade-style game about clouds and rain. Feasibly, it could even be worksheets and lectures … but why would you want it to be? This is the interesting part! Get creative!

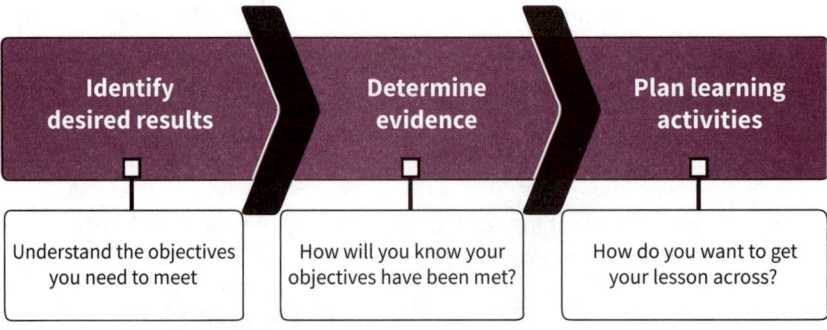

Identify desired results	Determine evidence	Plan learning activities
Understand the objectives you need to meet	How will you know your objectives have been met?	How do you want to get your lesson across?

Source: "Backward Design" graphic created by Kiki Prottsman ©2023 Kiki Prottsman.

Find Common Ground

While working on your lesson, look for common ground between the CS standards that you need to hit and the standards and objectives from your core class. It doesn't usually take too much scrutiny to find similarities.

Studying World War II in history class? Connect it to the *standard* on data representation and have students build code-breaking programs. Working with the scale in music class? Have students create songs in a coding program that can shift up or down the scale using the press of a button.

Remember, the key to efficient integration is doing more things in the same amount of time. Your core classes don't need to suffer when you add CS to the mix. One can truly enhance the other.

Let Students Guide You

If you worry that your new lessons will come off as disconnected, consider letting students come up with the integrated lesson content. Have students identify elements of computer science that they would like to see combined with your classes and have them dig into the ideas that need to be learned to make them happen. Then, have students teach one another what they've discovered. This could lead to some phenomenal explorations in cutting-edge topics like AI, e-textiles, alternate reality, sports performance systems, or cryptocurrency.

> *"The cure for boredom is curiosity. There is no cure for curiosity."*
> — *Dorothy Parker*

Test It for Fun

Sometimes we fall into a comfortable pattern of doing the same type of activities over and over for different lessons because it's comfortable and easy. Big corporate curriculum houses often publish lessons like this because teaching for a test feels efficient, and spending a day engulfed in play can feel frivolous ... especially in high school. But think back on the lessons from your childhood that stood out the most. Quite often you'll find that they were memorable because they were different.

Injecting a lesson with moments of pure fun will help students identify the responsible subject areas as interesting, priming students to pay more attention and achieve more.[27]

Summary

At first glance, the world of cross-curricular lesson plans may seem vast and barren. When you know where to look and how to modify lessons that already exist, the process of adding computer science to your core class is more than possible; it can be awesome.

No matter how you acquire lessons for your classroom, try to keep them as relevant and fun as possible. And, if all else fails, you can take a shot at creating integrated lessons, yourself!

Reflection Questions

1. This chapter gave some general tips on curriculum for computer science integration. Take a look at some of the prebaked lessons and curriculum sets that were shared to see if any of them would work for you. How many look like they'll work well right out of the box? How many could work with a little tweaking?

    _____

    _____

    _____

    _____

2. Sometimes, premade curriculum provides a small age range, either because of the way that it's written, because of the images included, or because of the standards that are targeted. See if you can find a lesson or two that is advertised for an age range other than what you teach. Can you think of any ways to differentiate it so that it works for your audience?

    _____

    _____

    _____

    _____

3. When writing a curriculum, getting started can be the hardest part. To avoid that hurdle, take a look at one of the lessons that you currently run in your classroom and compare that against the CS standards suggested by your district. Can you find a way to adapt your current lesson to incorporate any of the standards identified for your age group?

    _____

    _____

    _____

    _____

Integrating Computer Science Into STEM-Focused Classes

CHAPTER #5

Q: Where can I find quality resources for integrating computer science into my math or science classes?

A: Just a couple of years ago, this question would have been much more difficult to answer … and a few years from now, it will be much simpler. For now, you can visit your favorite search engine (bonus points if it's connected to an **AI** backend) and enter the prompt: *"grade _____ lesson plan for integrating computer science into _____."* You should receive several pages of suggestions for math or science. From there, take a look at the lessons to see if any are similar to one that you already do or teach. If you don't see specific standards alignments within the lessons that you choose, you may need to validate the activities against your local standards documents, but that should go faster than blending your own lesson from the beginning.

When it comes to the maths and sciences, technology feels like a natural fit. That's probably why there are entire curriculum sets dedicated to teaching math and/or science using CS concepts (like Bootstrap or ProjectGUTS, mentioned earlier). But what is a teacher to do when they're required by their district to teach a specific series and not deviate from what their neighboring classrooms might be doing? This is where we need to get creative … and it's also where the magic happens!

Step #1: Friday Fun Day

We all have those days when you know it's going to be tough to get the students invested. It might be the beginning of winter when the snow has just come back to town, or in the days leading up to a break. Whatever the occasion, consider skipping the "movie day" and introducing your students to some enticing computer science lessons instead.

There are loads of one-off coding activities on HourOfCode.com that take little to no teacher prep, and no previous knowledge for students. Most of these are intended to take only a single session and appeal to a wide audience. Just head over to https://hourofcode.com/learn and click on the "science" or "math" filters to get lessons that are relevant to you.

Step #2: Project/Presentation Replacement

If you already spend a good number of days on projects and presentations, consider encouraging students to complete those in software form. If they've had a few "Friday Fundays" to grow comfortable with a programming interface (like Scratch, MakeCode, or Repl.it) let them spend their project prep time using trial and error to create an illustrative game or utility, instead of burying themselves in slides and poster paper. With the proper resources, it doesn't take much for students to develop portfolio-worthy creations.

Source: Microsoft MakeCode.

Step #3: Truly Integrate Lessons

This is the gold standard for 21st century education—replacing the lessons that you used to offer with enhanced versions that are rich in computer science exposure. It may be a few years (or half a decade) until the large curriculum providers have high-quality solutions, so let's explore ways that we can do this ourselves.

Fortunately, there are several CS standards that naturally coordinate well with math and science, especially when it comes to collecting and representing data. Look for moments in your lessons where you can showcase different ways to present the same information. You can take this one step further by helping students to analyze their data to support a claim. Both of these are common occurrences in subjects like math and science. Collecting and exchanging data via digital means is another great integration point.

Find moments in your lessons where students can collect information, process that information using software, then present the information in multiple forms. Those simple lesson modifications hit multiple standards for Grades K–10 and require very little extra in the form of time or equipment for activities that already focus on information gathering or processing.

"Absorb what is useful, reject what is useless, add what is specifically your own."
— Bruce Lee

Beyond data collection and modeling, coding is a great way to help students digest the ideas they are learning. The beauty of writing code is that you need to understand the way things work to represent them computationally. This is part of the reason why computational thinking exploded onto the scene in the early 2010s as CS gained popularity.

Computational thinking is generally said to be made up of four pillars:

- Decomposition
- Pattern matching
- Abstraction
- Algorithms

Decomposition

Decomposition is the ability to break a big problem down into smaller pieces to make the issue more manageable. It's like pulling a story problem apart to know which steps need to happen to solve it.

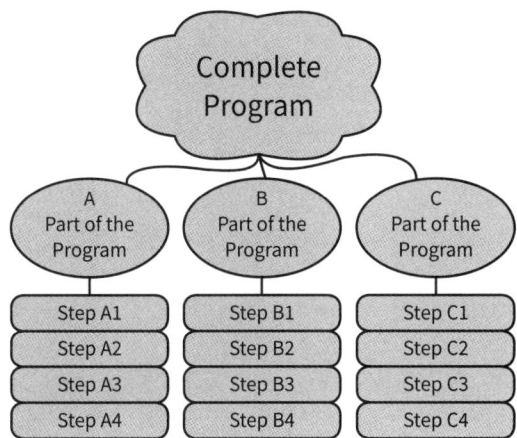

Pattern Matching

Pattern matching is the ability to look at a problem and see how it might be similar to something else that has been tackled before (or how it is similar to something else that needs to be tackled). It's like recognizing that a story problem on a test is just like a story problem assigned in homework, but with oranges instead of apples.

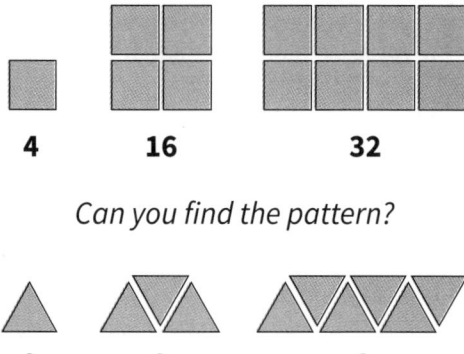

4   **16**   **32**

Can you find the pattern?

?   **?**   **?**

Abstraction

Abstraction is the ability to focus on what is important and abstract away little details. It's like realizing that it doesn't matter whether the girl in a story problem is named Tish, Latoya, or Betty … or whether she's buying apples, oranges, or beach balls; you can create a formula that will solve the problem for any of those cases.

Abstraction also refers to the idea that you don't always need to know every detail to count on a result. For example, in Figure 5.1, you don't need to know *how* the box does what it does. All you need to know is that one thing goes in, and two of them come out.

Figure 5.1   Magic Doubling Box

$$x \rightarrow \boxed{} \rightarrow 2x$$

Source: Kiki Prottsman, 2023. Elements designed using Adobe Illustrator with AI generation.

Algorithms

Algorithms are tools to reconstruct and relate the recipe for solving your problem in a consistent and meaningful way. It's like coming out the other side with a mathematical formula that you can use to solve the story problem (and maybe even use it to solve other problems in the future!).

Step #1: Lift coffee to mouth
Step #2: Drink coffee
Step #3: Set down coffee

Source: Kiki Prottsman, 2024. Elements designed using Adobe Illustrator with AI generation.

Putting the Pillars Together

When students know how to employ these pillars of computational thinking, they have what they need to break down and digest the ideas at the heart of any subject—whether that be the water cycle, the structure of long division, states of matter, or linear equations. Students will need to decompose the concept into base components, figure out how each piece relates to the application that they want to create, abstract away the parts of the topic that aren't critical to the assignment, and construct a project that meets requirements.

Coincidentally, that's also what teachers need to do to construct their own integrated lessons for these subjects. Imagine that you already have a lesson on two-step equations. Perhaps it consists of three days of material: some lecture time, some worksheets, some partner work—plus a few nights of homework. We can break those lessons down into their most important parts, find patterns between your favorite activities and popular activities that integrate CS, abstract away elements that aren't crucial for learning two-step equations, and weave together a functional and entertaining integrated lesson.

Decomposing Lessons

Let's give this example a try with one of your actual lesson activities. Grab a unit that you would normally teach and pick out the main objectives. Now, pick a computer science standard that you would like to hit.

For my example, I'll continue with two-step equations and my objectives will be:

▶ Use inverse operations to solve two-step equations

▶ Describe how operations must occur on both sides of the equals sign for the statement to remain true

▶ Verify potential solutions by substituting the answer for the variable

My computer science standard will be:

▶ CSTA 2-AP-16: Incorporate existing code, media, and libraries into original programs, and give attribution

Notice that this standard does not dictate that computer science be the *focus* of the lesson; it only suggests that students need to understand that attribution needs to be paid when using other people's assets. This is perfect for my purposes, because I can prepare a series of puzzle pieces for students to choose from when trying to create an equation solver, and that will cut down our computer time significantly.

Perfect. Now I have my lesson decomposed into the pieces that I need students to learn, and I'm using backward design to boot!

Pattern Matching Lessons

This is a great time to go through lessons for other integrated topics and see how the creators weaved CS into another subject. Did they have students fix broken programs that became utilities for their assignment? Did they encourage students to work together to brainstorm ways to depict story problems using choreographed sprites? Did they suggest that students use their own encoding of secret messages to transport data from point A to point B? What can you find that you can tailor to the lesson that you're building right now?

As I noted earlier, I'm going to follow the model of providing disjointed chunks of code that students need to rearrange via logic and reasoning to create a two-step equation solver. With any luck, the process of figuring out how to arrange prewritten code will help internalize why we use inverse operations to simplify an equation, as well as the usefulness of keeping an equation balanced by performing operations on both sides of the equal sign. Since this was predicated on the ideas in CSTA Standard 2-AP-16, I am confident that objective will be met, as well.

Abstraction Across Lessons

In our case, this step aligns tightly with pattern matching. To clearly see the patterns between the lessons that others have done and what we want to do, we'll need to ignore the fact that the inspiration lesson may have been about the lifecycle of a butterfly, and instead extract reusable core takeaways.

For my purposes, I'll be ignoring the exact subject of the inspiration lesson, as well as the actual content of the previous code and the language in which the code was written. All those things can be replaced without losing the standards and objectives that I'm trying to hit. And I'm not worried about the fact that I still have one objective left, because this section is about integrating CS lessons, not about completely displacing traditional teaching. I can still use worksheets and videos to pick up additional ideas.

Algorithmifying Your Lesson

Ignoring the fact that it's not a real word, you can finish this process by algorithmifying your lesson. This is just to say that you'll write it up into a lesson plan that you can follow when you're ready to run this with your classroom. If it's successful, you can probably even sell it on Teachers Pay Teachers for $2–$12 a pop and be a hero to educators just like yourself! If it's not as successful as you would like, make notes on your plan and tweak it for next year.

Play Time

Math Game Time

Do you want to see the classroom activity that I put together to help students understand two-step equations? Head over to aka.ms/two-step to see if you can get the program working. Don't forget to give proper attribution when you share your finished program!

When you start looking for moments where you can replace past go-tos with CS integrations, you'll start finding them everywhere. Whether it's Friday Fun Days, project and presentation swaps, or elements of daily lessons, you can fit quite a bit of computer science into your daily subjects without losing any valuable instruction time. In fact, you might find that the project-based, constructionist, exploratory nature of computer science helps students to better internalize concepts that they merely memorized before.

Spotlight

Interview With a Colleague: Maria Sellers

Maria Sellers is a seasoned educator in Indiana who is changing the way computer science is taught in local schools. With a rich background in music education and e-learning, Maria spent years perfecting the art of curriculum design. Recently, she teamed up with Nextech to cook up lessons that mix computer science into everyday math and science classes.

Maria starts by finding common ground among subjects. She starts by looking for standards that align, linking existing CS lessons onto the end of math or science lessons that share topics or objectives. She considers both "plugged" and "unplugged" options when creating her combinations, offering teachers a variety of ways to introduce concepts into their classrooms, providing a complementary and enriching experience.

None of this work is done in isolation. Before proposing lessons, Maria consults with teachers to understand their objectives and the intentions behind their existing activities. Using this information, she crafts lesson combinations that align with the original vision of each class. She recognizes that it can take an educator several years of working with a subject before they become comfortable freely customizing lessons for their students, so she strives to create experiences that are relevant and authentic.

Maria firmly believes that computer science is as fundamental for future generations as reading and writing. She jokingly asks, "Do you want to be smarter than your smart home or do you want your smart home to outsmart you?" This highlights her argument that a foundational understanding of computer science is essential not just for daily interactions with technology but also for ethical and societal implications, such as preventing the misuse of artificial intelligence. This is becoming increasingly important!

Maria believes that one of the largest misconceptions held by teachers is that they need to be experts in computer science before they can begin to teach it. She counters this by emphasizing that computer science encourages learning through failure rather than rote memorization, and as long as instructors remain one lesson ahead of students, they'll be in a good position to teach the material.

Over the last decade, Maria has noticed a significant shift thanks to CS integration in elementary schools. Students used to enter her computer science classes in junior high with looks of fear and confusion when confronted with words like *algorithm* and *conditionals*. Now, they arrive with a better understanding of the topics and are more excited than scared to tackle her assignments. Her classroom, once believed to be an option reserved exclusively for "the smart kids," has become accessible and welcoming to everyone.

Interestingly, Maria pointed out that every computer science class at her school in western Indiana is now led by a female educator, challenging stereotypes and inspiring the next generation.

Educator Takeaway: When you use backward design, it's possible to create meaningful CS experiences for students in any class—and even small amounts of computer science in early grades make a difference later on.

Summary

You may remember computational thinking as a trendy subject from the last decade, but decomposition, pattern matching, abstraction, and algorithms can help you create integrated lessons that stand out. If you pay attention to your district's standards and backward design, you can create fun and effective activities that allow you to stay true to your subject without adding extra planning time term after term.

For schools without strict CS integration standards, you can provide a satisfying taste of CS with one-off activities during free time, unit breaks, and holidays.

Reflection Questions

1. Without regard for practicality or your current level of training, list five separate lesson ideas you would like to explore when it comes to integrating CS into your math or science class.

   _____

   _____

   _____

   _____

2. Take a look at your list from Question 1. Which exercise looks most like something you already plan to do with your class? How much CS would you need to add to your activity to meet one or more of your district's CS standards?

   _____

   _____

   _____

   _____

3. What preparation would you personally need to write yourself a complete guide for the lesson you chose? Can you identify any resources from the previous chapters in this book that could help you feel prepared? Write down two resources you are willing to look into further to complete your custom lesson by the time the next school year is over.

   _____

   _____

   _____

   _____

CHAPTER #6

Making Any Class a STEM Class Through Coding

Q: Do I really need to try to integrate computer science into my class if I don't teach in a STEM area? What does computer science have to do with art, music, or PE? I'm worried these lessons would feel inauthentic as part of my curriculum.

A: Quality computer science integrations may feel farther away when you teach a non-STEM class. But remember, we're trying to meet students where they are. If we want all students to have a chance at participating in the jobs of the future, we need to be able to reach students no matter where they consider their strengths to be. It is possible to add high-quality CS experiences to all sorts of creative classes without taking away from your core learning objectives.

Don't worry about becoming an expert in coding or hardware—instead, think about technologies that already accompany your field, and consider the ways they might affect people from different groups or the ways they bring joy to those who use them.

As you read through this chapter, keep in mind that you don't need to shift focus away from your main subject to incorporate a decent amount of computer science. So much of CS translates between topics, and multiple standards can be explored with relatively small modifications. Don't worry about becoming an expert in coding or hardware—instead, think about technologies that already accompany your field, and consider the ways they might affect people from different groups or the ways they bring joy to those who use them.

By now, you'll probably be able to follow the steps in the previous chapter to create your own custom integrations, but I'll also take some time to paint a better picture of how CS education can be specifically included in non-STEM classes. Exact methods may change from subject to subject, so I'll peel ideas out by category—but it's a good idea to browse all sections, as some recommendations cross over quite well.

ELA/History/Social Studies/Foreign Languages

These classes often revolve around stories: writing stories, reading stories, and understanding the stories of others. The following exercises take that into consideration.

Simple Additions

1. **Coding Stories:** Several block-based coding platforms allow students to create story-based projects. Code.org has Sprite Lab, MakeCode has Arcade, and MIT has Scratch. All of these platforms allow students to design visual stories that can be used to integrate computer science through the use of basic coding concepts.

    Imagine having students create projects that represent the events of Victorian England or early civilization using "sequences, events, loops, and conditionals" (CSTA Standard 1B-AP-10).

2. **Anachronisms:** Connect the present to the past or future through hypothetical exercises. Without the use of any extra technology, you can achieve standards for multiple grade levels through the "Impacts of Computing" track. Add a moment in your unit that requires students to empathize with a character within the world you're studying and ask what technology might look like where they are.

    Imagine having one student pretend to be Fern Arable from *Charlotte's Web* and another student pretend to be a reporter. Through interview questions, the two could "Compare tradeoffs associated with computing technologies that affect people's everyday activities and career options" (CSTA Standard 2-1C-20) in the context of how Fern's life might have been different if Instagram and YouTube had been around during the events of the book.

Deeper Integrations

1. **Build an App:** Challenge students to solve a problem found in the time and location of your unit. Make sure that students know their final project will be presented in slides as a screen-by-screen breakdown of what their app would be if they had been able to program it for real (if your school has a consistent base of CS across the board, feel free to request that students *do* program it).

    Letting students know in advance that this will be a required project will help them keep a lookout for moments that can be captured within their solutions. This project will help students discuss the pros and cons of technology and practice creating pseudocode that can be clearly relayed as part of a presentation.

    Imagine if, as students learn their Spanish nouns and verbs, they know they'll be responsible for proposing an app at the end of the term that solves a problem for the people of Spain. Throughout the term, they can take moments to "brainstorm ways to improve the accessibility and usability of technology products for the diverse needs and wants of users" (CSTA 1B-1C-19), "use flowcharts and/or pseudocode to address complex problems as algorithms" (CSTA 2-AP-10), and "seek and incorporate feedback from team members and users to refine a solution that meets user needs" (CSTA 2-AP-15).

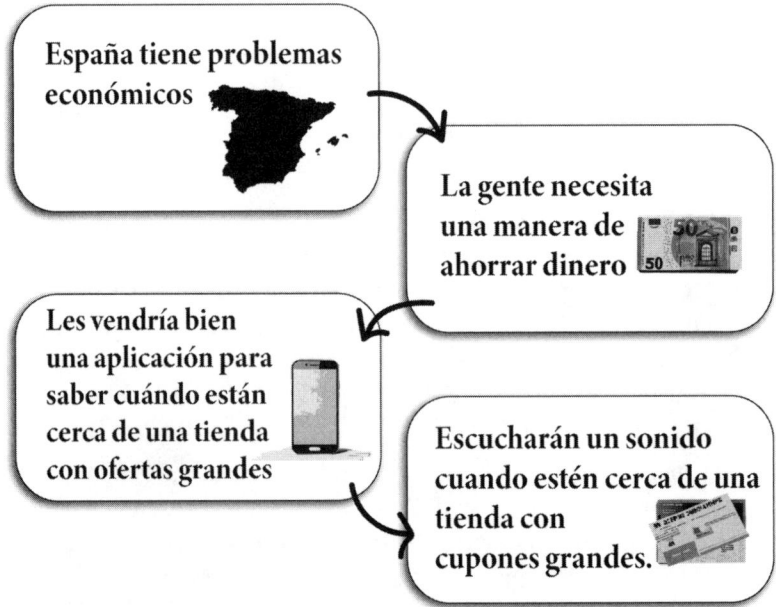

Source: Kiki Prottsman, 2024. Elements designed using Adobe Illustrator with AI generation.

Art

Art is critical to technology and technology can inspire art. With AI blasting onto the scene in recent years, artists may feel like computers are the enemy when it comes to making a living in the space. However, when used responsibly, AI can help artists overcome major hurdles, including artistic blockage and monotony.

Beyond AI, the generation of digital art can be accomplished by programs that realize mathematical patterns or interactive transformations. Though the results may be displayed on a screen or transferred to a canvas by robotic parts, the artist is the person who dreamed up the concept and performed the work that caused it to move out of the imagination and into the world—and therefore the intellectual property belongs to the artist.

Simple Additions

1. **Art in Games:** There are a handful of block-based editors that allow students to create art as part of larger projects. MakeCode has Arcade, MIT has Scratch, and Code.org has Sprite Lab.

   Imagine having students work together to create animated vignettes in MakeCode Arcade while they have been asked to "distribute tasks and maintain a project timeline when collaboratively developing computational artifacts" (CSTA Standard 2-AP-18).

2. **Mathematical Art:** There are a few programs at the beginner level that allow students to directly craft artistic creations using code. Some of these include Art by Code.org, Code Maven by Crunchzilla, and Intro to CS by Khan Academy.

   Imagine practicing color theory using Javascript and assigning students the task of creating a program that allows them to input two colors and receive the output of what they would create if mixed. If, in the process, they are assured to "create clearly named variables that represent different data types and perform operations on their values," then they would experience CSTA Standard 2-AP-11.

   For younger students, you could run the activity in a similar way, but instead of coding in JavaScript, have students write pseudocode for one another (relying on variables to hold the inputs) and have the receiving student manually mix the color and return a swatch with the correct color.

3. **Minecraft Education:** Game designer Mark Rosewater said "restrictions breed creativity" and John C. McCrae wrote "limitations foster creativity" within artists. What better way to honor that overlap than to create art within a game? Minecraft Education Edition allows students to spend time in a place they likely already know, confined by blocks and squares but inspired by structures and ideas from the outside world. In the educational version of the popular game, students will find they're accompanied by a robot (called "the Agent") who can follow their directions to quickly build nearly anything they could make "by hand."

   Imagine assigning a project where students are to replicate famous sculptures using the Agent within Minecraft, while requiring that they "document programs [using comments] to make them easier to follow, test, and debug" (CSTA Standard 2-AP-19).

   For younger students, you could allow them to create their own small sculptures freeform using the same tools.

Play Time

Give the Education Edition of Minecraft a try and see how easy it is to get started with the platform. Visit https://bit.ly/3QCtCE2 to navigate to their free sample activity and follow the directions.

Bonus points if you play alongside a younger family member or a student. What is their reaction to Minecraft Education?

1. **AI:** Students can now utilize free **LLMs (Large Language Models)** such as Dall-E to generate initial drafts of art concepts they will be able to expand upon using the techniques provided to them in class.

    Imagine assigning students a painting survey where they need to emulate the style of Claude Monet, but with a subject near and dear to their own lives. You could instruct students to prompt an LLM of their choice with the details they want it to consider, then require them to print out one of the resulting images as a point of inspiration before creating their own painting. In the process, if you ask them to "describe tradeoffs between allowing information to be public and keeping information private and secure," you will have incorporated CSTA Standard 2-IC-23.

    For younger students, you could run the activity in a similar way, but have another student act as the "AI" in this exercise.

Deeper Integrations

1. **Digital Portfolios:** There are a plethora of apps online right now where students over the age of 13 can document the art

Source: Kiki Prottsman, 2024. Created using Midjourney and edited with graphic design programs.

they have created and describe techniques they used within their process. Through Instagram, Tumblr, Behance, or Flickr (or the photo app of your choice), students can post images and leave comments with notes about their methods and inspiration.

Imagine assigning students to document each project they create as they work, with a wrap-up post at the end. Pair this with hefty doses of internet safety and cybersecurity discussions throughout the term and you could potentially touch on standards (see Table 6.1).

Table 6.1   Example Standards Pulled From CSTA in 2023

1A-1C-18	Keep login information private, and log off of devices appropriately.
1B-NI-05	Discuss real-world cybersecurity problems and how personal information can be protected.
2-NI-05	Explain how physical and digital security measures protect electronic information.

For younger students, hit the same points with your own "online" social app created by stringing a piece of twine from one wall to the other and asking students to paste their art to a sheet of construction paper before writing project details on the back. Clip this to the line to display throughout the term. Additional pieces can be added beneath to create their very own thread.

Source: Kiki Prottsman, 2024. Created using Midjourney and edited with graphic design programs.

Older students with more coding knowledge can create and curate their own webpages for their art, getting exposure to an even larger number of standards in the process.

Music

Artificial intelligence has not come as far with music as it has with graphic arts and writing, but breakthroughs are just around the corner. Nevertheless, technology has been well integrated into music for decades. Whether it be the synthy-pop sounds of the '80s and '90s or the newest DJ mixing equipment, it's hard to imagine any music (live or recorded) without a little bit of digital magic.

Simple Additions

1. **Music for Games:** When it comes to adding music to larger game worlds, MakeCode Arcade has come a long way beyond what its peers can handle. With a full music staff, including bass clef, an adjustable number of measures, tempo changes, and the ability to modify time signature, students can create background music, sound effects, and themes with very little prior experience. Close behind is MIT with Scratch, allowing students to compose their own music with individual notes played by an impressive number of instruments at variable tempos.

   Imagine providing students with a premade scene in MakeCode Arcade and asking them to work in teams to compose background music and sound effects to go with the events that you've constructed and embed them into the correct places in the program. You may even have them "take on varying roles, with teacher guidance, when collaborating with peers during the design, implementation, and review stages of program development" (CSTA Standard 1B-AP-16).

2. **Looping Beats:** More intermediate students can bring text-based coding into the equation, using applications like EarSketch or Sonic Pi. Either of these will allow you to use code to compose your own creations on the fly.

   Imagine assigning students a project where they need to explore chord progressions and keys while attempting to create harmonies and dissonance. Ask them to record data on their experiments and display their results in a way that helps them gain deeper understanding.

Deeper Integrations

1. **Microcontrolled Music Player:** Consider helping students construct their own instruments or music mixers with inexpensive hardware. Small, low-cost microcontrollers

are coming back in stock. As a result, you can easily get the ingredients to create custom music using the micro:bit, Makey Makey, Raspberry Pi, or Arduino.

Imagine using the micro:bit to hold simple songs and allow students to launch them with the push of a button. As their knowledge expands throughout the year, you can add harmonies, chords, and beats to the mix using other available events and triggers. If students already know some coding from other classes, or if they're able to pick it up quickly throughout the year, they have the potential to hit standards (see Table 6.2).

Table 6.2   Example Standards Pulled From CSTA in 2023

2-AP-17	Systematically test and refine programs using a range of test cases.
3A-AP-17	Decompose problems into smaller components through systematic analysis, using constructs such as procedures, modules, and/or objects.
3A-CS-03	Develop guidelines that convey systematic troubleshooting strategies that others can use to identify and fix errors.

Physical Education

Are you surprised to find physical education in the mix? Integrating CS into PE can offer enrichment opportunities for students that can help students who don't like PE to like it better and likewise encourage students who don't like CS to like it better.

Simple Additions

1. **Unplugged:** Use simple, unplugged lessons as games to encourage computational thinking while engaging in physical activity. Some great lessons can be found on Code.org or CSUnplugged.org.

   Imagine using the Relay Programming activity from Code. org. Students must use a special code language to describe the drawing of a picture on a grid. Each student runs to the other side of the field to add their step to the program, making sure there are no bugs up to that point. The first team to correctly finish their program wins. Make sure that each player takes time to mentally "test and debug a program or algorithm to ensure it runs as intended" (CSTA Standard 1B-AP-15).

2. **Logging Info:** Put kids in teams to run races, bounce balls, hit shuttlecocks, or throw discs. Have them collect data on the activities and present those data to determine where their stats fall within the achievements of the class.

Imagine having a field day where students must keep track of scores on a tablet. At the end of the event (or during the next meeting) have them combine and sort the data to determine which team came out on top in each activity. By the time the ribbons are awarded, they will have had to work together to "collect data using computational tools and transform the data to make it more useful and reliable" (CSTA Standard 2-DA-08).

Deeper Integrations

1. **Tossing Tech:** If you want an inexpensive and mobile data collection unit, that's where the micro:bit really shines. With microphone, accelerometer, compass, and temperature sensor, students can conceive of and utilize this small piece of equipment to gather more details on physical adventures. Pair up with the STEM class to have the devices programmed with the code that your students devised.

   Imagine taping your micro:bit to the top of a frisbee and gathering data with each throw. Students can analyze compass heading, light level, sound level, rotation, and acceleration and see if any of these factors correlate with an increase or decrease in overall distance. Bring the invention out several times a year to gather data on improvement during the beginning, middle, and end of the term. If students design and build this contraption on their own, they could potentially be exposed to the standards that follow (see Table 6.3).

Table 6.3   Example Standards Pulled From CSTA in 2023

2-CS-01	Recommend improvements to the design of computing devices, based on an analysis of how users interact with the devices.
2-CS-02	Design projects that combine hardware and software components to collect and exchange data.
2-CS-03	Systematically identify and fix problems with computing devices and their components.

If your class lands outside of these defined spaces, don't despair! Just take a good look at what I did to weave together typical subject lessons with the most convenient CSTA Standards and do something similar with your class. Need to hit standards that are more complex or advanced? Any combination is achievable as long as you are willing to explore deeper possibilities and lean on your network for advice and lesson reviews.

Spotlight

Interview With a Colleague: Holly Swartz

In a world where technology often takes center stage, Holly Swartz has found a unique way to integrate computer science into her art classes, proving that the two subjects are not mutually exclusive but rather complementary avenues for creative expression and problem-solving.

After 19 years of teaching in an elementary classroom and 5 years as a digital integration specialist, Holly moved school systems with the hopes of teaching STEAM (Science, Technology, Engineering, Arts, and Mathematics) but found herself leading an art class. Undeterred, she decided to subtly infuse her lessons with computer science, introducing pixels as a bridge between video games and Pointillism. She also incorporated Spheros into her class, encouraging students to program the robots to navigate through paint, creating intricate patterns and color combinations.

Holly's teaching methods extend beyond traditional art mediums. She sometimes employs Google Slides when teaching her students about binary language, then she encourages them to encode their names and deliver personalized presentations. One of her activities involves students coding their friends to get them to draw a specific, predetermined image. The exercise challenges students to think algorithmically, providing step-by-step instructions to recreate a picture they can't see.

One holiday season, in a playful response to a STEAM teacher who usurped her classic "hand turkey" art project, Holly decided to have students code an algorithm for drawing a turkey created from hand shapes, instead. Holly also enjoys providing other holiday-themed integrations, like challenging students to recreate a specific Christmas tree from a list of steps before allowing them to add their own finishing touches.

For younger students, Holly uses MakeCode Arcade, enabling them to see their art come alive instantly as they build games. Older students are introduced to Artist from Code.org, where they can create more complex and intricate images.

Holly believes that teachers don't always need to make a concerted effort to bring a large amount of computer science into the classroom. It can be enough just to start using CS vocabulary words in early education. She encourages teachers to pull computer science ideals from children's books during story time, pointing out that teachers don't need to use books written specifically for CS for this to work. Concepts like repetition, conditionals, and sequences show up in lots of books for young readers.

According to Holly, integrating computer science is not just about teaching kids to code; it's a tool for problem-solving, learning to iterate across ideas, and persisting without receiving answers directly. She often shares the quote from Thomas Edison: "I have not failed. I've just found 10,000 ways that won't work," believing that this way of thinking can help students in all subject areas as they grow.

Holly's teaching philosophy rejects the traditional "talking head" approach; she gave up her teacher's desk years ago. Instead, she prefers a dynamic environment where she can engage with each student and encourage them to think independently.

The impact of Holly's integrated approach can be felt long after students leave her class. Students return from time to time to credit her innovative teaching for their successes, citing reduced test anxiety and increased courage to pursue their dreams. In blending art and computer science, Holly Swartz has not only enriched her curriculum but also empowered her students to see possibilities that arise when fields intersect.

Educator Takeaway: Computer science lessons are often seen as "fun time" when integrated into other classes. You might be surprised what even your youngest students are capable of when they believe in their abilities, and sometimes those memories shape the rest of their lives.

Summary

While it might not seem like computer science blends well with the subject that you already teach, technology is everywhere. If you and your students dig into the tech that makes your field shine, you'll realize that there is likely some kind of hardware, software, or theory that can be studied to better understand how CS affects the craft. Not only is it possible to create authentic lessons that integrate CS standards, but it can also be enriching and fascinating.

Reflection Questions

1. Reflect on the broader theme of custom CS integration discussed in this chapter. What stood out as a strategy that could be applied in your specific subject area?

    _____

    _____

    _____

    _____

2. Suppose that your school was planning to celebrate "Art Day." Can you conceive of any special projects that you would be willing to integrate into your lesson in celebration of that event? Now imagine that it was "Digital Art Day." Can you modify the project or activity that came to mind in such a way that students could generate art via computational artifacts while remaining on-topic in your classroom?

_____

_____

_____

_____

3. Are there any other teachers in your school who plan to bring computer science into their classroom? Could you see yourself working with one or two colleagues to come up with a project that could spread across all of your subjects? How might students plan in one class, code in another, and test in a third? What kind of lesson could you come up with if you knew you would have another educator to help implement the computer science content?

_____

_____

_____

_____

4. What do you think you would need to learn if you were going to be the implementing educator identified in the previous question?

_____

_____

_____

_____

PART 3
TEACHING COMPUTER SCIENCE FOR TODAY AND FOR THE FUTURE

We all have our strengths.

"Everybody is a genius. But if you judge a fish by its ability to climb a tree, it will live its whole life believing that it is stupid."

—Unknown (often attributed to Albert Einstein)

CHAPTER #1

Supporting Computer Science Students From Historically Underrepresented Groups

Q: I really want to encourage students from historically underrepresented groups to take my courses, but I see the same demographic year after year. What can I do to change that?

A: Remember that students talk! Regardless of the way you advertise your class, students will share their experiences—and in the same way that students know which teachers are coveted and which they should avoid, they'll know which classes help them feel empowered and which make them feel out of place. Equity should be an everyday objective.

We touched on the importance of equity and inclusion in Chapter 2. Here, we'll dive deeper into the ways that diversity improves teams, product, and profit. We will also explore techniques that you can use to ensure that your classes provide a welcoming environment to as many students as possible.

Diversity in computer science isn't meant to be a token gesture; it's foundational to the field, driving creativity, innovation, ethics, and global reach. It ensures that technology serves the needs of the entire population. By embracing diversity, the CS community can contribute to a more equitable, innovative, and socially responsible world. That means recruiting and retaining students who are less likely to opt-in to computer science classes (aka CS students who are from historically underrepresented groups). Generally, students from historically underrepresented groups fall into a wide range of categories: nonmales, Black people, Indigenous people, Hispanic people, people from the LGBTQIA+ community, individuals from low-income backgrounds, people who gravitate toward sports or humanities . . . all of these groups are less likely than others to choose computer science when offered as an elective.

Let's step back and take a look at what happens when organizations don't intentionally seek diversity during the development of technological products.

Bias in Voice Assistants and AI

How many of you have voice assistants (Alexa, Google Home, Siri, etc.) in your household? If you have a feminine voice, you probably already know where I'm going with this. For everyone else, you may have no idea what the problem is. I'll enlighten you.

Statistically speaking, voice assistants have a much harder time understanding feminine voices than the voices of their masculine counterparts, failing women nearly 21% of the time.[28] I can't even tell you how frequently I've had Alexa ignore or misunderstand me when I'm trying to ask for advice with my hands full, just to have my adult son repeat the exact same query and receive an answer. It's frustrating! It feels a lot like what a woman experiences in the boardroom, except that it's happening in my own home.

According to Dr. Joan Palmiter Bajorek, voice assistants struggle to parse "breathier and higher-pitched voices." This isn't a new revelation, either. This has been a known issue since the days of interactive car systems, when the auto industry recognized that female drivers were more frequently misunderstood.

Their proposed solution? Teach women to speak more like men.[29]

So, why is it that AI has such an affinity for male voices? It's because the teams that developed the technology were made up almost exclusively of men. This is a common problem with technological products due to the fact that computer science is nearly 80% male-identifying,[30] and though it's said that TED Talks are frequently used for speech analysis, a homogenous group of men is much less likely to notice that 70% of TED Talks are given by men. Adding even a single woman or nonbinary member to such a group brings in a new perspective and increases vigilance.

If we had more women involved in decision-making during the technology creation process, we might not be in a situation where nearly all digital personal assistants are represented as females while nearly all law and finance bots are represented as males.[31]

> *Nearly all digital personal assistants are represented as females while nearly all law and finance bots are represented as males.*

For better or for worse, human biases transfer into artificial intelligence, and while this probably won't be eliminated as long as AI is trained on human data, it can be mitigated with more checks and balances in the form of diverse and diligent teams. Such teams will only become more important as more of our daily life is automated. From creative idea generation within ChatGPT to AI assistant hiring pipelines, these trained models have an increasing amount of influence on business and community; therefore, we are obligated to make sure they represent the society that we need and are not an amplification of the society we already have.

Spotlight

From Kiki's Life

Recently, I went to a workshop titled "ChatGPT and **Generative AI** in the Classroom," given by Jeff Utecht of Shifting Schools. One of the exercises Jeff had us participate in was to ask ChatGPT to craft a short story about a man and use the same prompt to craft a story about a woman. In other words, same story; only the gender changed.

In my male story, Mr. Jameson was described as "a mentor, a guide, and a hero." Mr. Jameson was a computer science teacher whose school was facing budget cuts, so he "devised a plan. He decided to host a Virtual World Fair, inviting parents, community members, and local businesses to experience the students' creations. He believed that if others could see what his students were capable of, they would understand the value of computer science education." Ultimately, "the success of the fair was overwhelming. Donations poured in, and local businesses offered support. The computer science program was not only saved but expanded, thanks to Mr. Jameson's unwavering commitment to his students."

In my female story, Margaret Thompson was "a mentor, a friend, and sometimes even a mother figure." One day, the school's heater broke, so "Margaret sent a message to the administration." In the meantime, to keep their minds off the cold, "she gathered all the children around her desk and began to tell them a story. Her voice, warm and soothing, filled the room as she spun a tale of adventure, courage, and friendship."

How many differences did you pick up on between these two stories? The male teacher is referred to as Mr. Jameson, while the female teacher is called Margaret. The male is quick thinking and business minded. The female is motherly and tells stories. The male is a problem-solver, he makes things happen. The female sends a note for help. The male attends to the educational needs of the kids. The female attends to the emotional needs of the kids.

These differences weren't accidental. They were inspired by billions of datapoints depicting centuries of historical prejudices. If we plan to turn this around, we can't rely on the systems we have now. We have to put a larger variety of people at the helm of the ship and work together to right our course.

Educator Takeaway: AI results are a reflection of the people who create and populate the system. We need to be conscious of this when we use AI to generate content. Also, we can help make this better by inspiring students from our communities to get involved with tech.

Bias in Face Recognition Software

Let's take a look at some of the ways that a lack of diversity in computer science has managed to cause true harm to communities of color.

In 2019, the National Institute of Standards and Technology (NIST) conducted a study testing 189 face recognition algorithms for accuracy and found "empirical evidence for the existence of demographic differentials in the majority of the face recognition algorithms."[32] Of the set of algorithms used to automatically detect whether the face in question matches a source image, NIST found that the overwhelming majority failed to give accurate results on certain members of the population. In fact, these algorithms are up to 100 times more likely to incorrectly identify Black and Indigenous American people than they are to incorrectly identify a white person. This could be a safety concern when it comes to allowing someone else to unlock a user's phone, but it could be a matter of life and death when it comes to falsely declaring someone to be a criminal.

> *These algorithms are up to 100 times more likely to incorrectly identify Black and Indigenous American people than they are to incorrectly identify a white person.*

Alonzo Sawyer spent 9 days in jail after facial recognition software mistook him for a perpetrator who was 7 inches shorter and 20 years younger. Nijeer Parks spent 11 days in jail after being mistaken for someone who left the scene of a crime after attempting to hit a police officer with his car (even though he had proof that he was 30 miles away at the time). Robert Williams was arrested for stealing five luxury watches, and when pointing out that the picture of the suspect looked nothing like him, the officer replied, "the computer says it's you."[33] All of these individuals are Black, and all were arrested solely on the premise that a computer had matched their faces to images of suspects.

Beyond race and color, automated algorithms have been found to unfairly target people with disabilities. Students with disabilities are more likely to be flagged by remote proctoring systems when taking exams due to disability-related modifications and affordances.[34] Similarly, Aggression Detection Technology, often used in schools to survey students as they enter the building, can disproportionally flag students who are hard of hearing, students who have trouble modulating their volume, and students with cerebral palsy.

Similar faulty algorithms have been used in surveillance software, airport screening devices, and even decision-making around jobs and housing. These systems, created primarily by able white men, continue to oppress and degrade others due in part to the lack of representation in the development process, and that's an injustice that has to be rectified immediately. We can't afford to continue bringing computer science only to students who are willing to opt-in without prompting. We need to reach out to all demographics at all ages and fight to create more diverse CS classes, particularly where Indigenous Americans, Black people, and people with disabilities are concerned.

Success Begins With Diversity

There are other benefits to cultivating diverse classrooms where many different types of people work together to bring projects to life. Consider how such an environment contributes to the belief system of your students. For example, suppose a boy starts his CS journey in a high school class with few women and very few people of color. Then, he goes on to college where he's in CS classes with few women and very few people of color. Next, his first working cohorts have few women and very few people of color. As he rises in rank, it's likely that he will have unconsciously developed the bias that the "right" demographic for people in CS is white and male. As a result, when he begins interviewing candidates for jobs, he may wholeheartedly believe that white male candidates come across as more qualified, whether or not that is objectively the case.

Scenarios like this have been playing out across the United States for decades—so much so that in 2014, Silicon Valley tech companies pledged to release diversity reports as a way of holding themselves accountable. That technique alone does not appear to be working, as the percentage of women in tech companies has dropped from 36% in 2014 to 34.4% in 2022 and the percentage of Black people dropped from 7.4% to 7%.[35,36] (Note that these demographics represent those employed at tech companies in general. If you remove participants working in human resources, customer service, etc., those numbers drop to a fraction of what is reported here.)

The United States isn't alone in this. Women make up only 22% of the tech workforce in Europe and 23% in Canada.[37] The percentage of Black people working in tech in Europe is just 3%[38] and 2.6% in Canada.[39]

This phenomenon isn't only doing a disservice to the individuals in question, it's hurting the bottom line of homogenous companies. Statistically speaking, organizations in the top quartile for ethnic, racial, and gender diversity realize larger financial gains than industry standard.[40] Diverse groups also consistently provide checks and balances for one another, allowing teams to make better choices and solve problems more effectively.

What's more, Millennials and Gen Zers tend to research company culture before applying for jobs, and more than three-quarters of job seekers believe that diversity and inclusion are important. A third of job seekers would not apply for a position with a company that lacked diversity (see Figure 7.1).[41]

Figure 7.1   Percentage of Job Seekers That Won't Apply Where Diversity Is Lacking

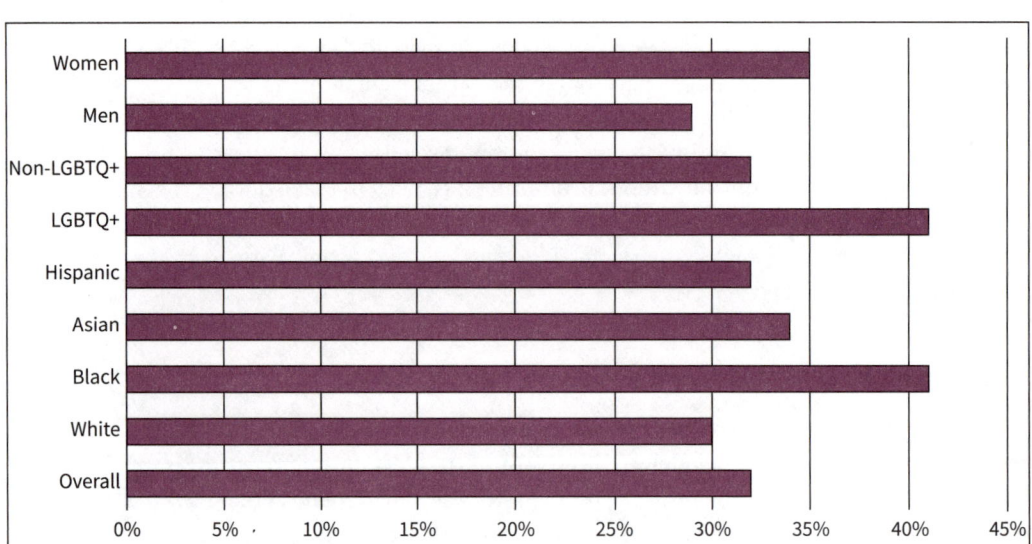

Source: Glassdoor, 2021.

Computer Science Drives Success

Stepping away from the big picture and major societal impacts of diversity in computer science, let's focus on the local impact that computer science can have on the life of a single individual.

Studies have linked computer science education to growth in creative thinking abilities, math skills, spatial skills, and reasoning.[42] Additionally, students allowed to work with others showed an increase in problem-solving capabilities. All these skills can help students excel in future classes, therefore boosting their employability, no matter what field they decide to pursue.

Additionally, a study conducted by Outlier Research & Evaluation out of the University of Chicago found that schools in Broward County who introduced a small amount of computer science into their fifth-grade school day saw significantly higher scores on the Florida ELA Assessment tests than those who did not utilize coding lessons. The score differential was even greater when teachers associated with the coding classrooms identified themselves as fluent in the skills of "resourcefulness and coping."[43]

Teaching computer science in elementary school has the added benefit of reaching neurodiverse students during their early years, while kids' minds are still most flexible and open to new languages and grammars.[44] Since autistic students and those with ADHD often find they excel at computer science in ways in which they don't in other areas, they can achieve early academic wins to help them develop a stronger sense of self-efficacy. When students have high self-efficacy, they tend to enjoy a challenge, persist longer, and be more resilient—all of which are traits associated with the ability to learn new skills.[45]

Hopefully, by now you see that computer science is not just another course that the administration is trying to cram into the school day. It's a pathway to strength, a skill for the future, and a mighty equalizer helping to ensure that all people can benefit from the products and tools shaping society.

With that, let's dive into some methods we can use in our classroom to provide the most welcoming and inclusive environment to students from all demographics.

Gaining and Retaining Students From Historically Underrepresented Groups

Getting students in the front door can be hard enough, but keeping them in the room and inspiring positive feedback is even harder! In this section, I'll share some of my tried-and-true methods for creating inclusive spaces, as well as strategies identified via research.

Reflect Your Community in Your Classroom

To cultivate a classroom that's welcoming to everyone, you will likely need to go out of your way to locate or highlight elements from outside your own culture—but there's a trick to doing that well. Be careful not to fall into the trap of believing that you know exactly what posters,

"We all should know that diversity makes for a rich tapestry, and we must understand that all the threads of the tapestry are equal in value no matter what their color."
— Maya Angelou

decorations, and music will speak to each group. You could end up perpetuating stereotypes or come off as pandering and inauthentic. James Charlton wrote a book in the '90s that leaned into the phrase, "Nothing about us, without us." No matter how much of an ally you consider yourself to be, I recommend memorizing that phrase and taking it to heart. Include your students in the process of preparing and decorating your space. Not only will the result be more representative of your students, you'll show them that your classroom is their classroom, too.

Encourage Mentors and Role Models

As with other STEM subjects, the presence of diverse role models and mentors can inspire and encourage students from underrepresented backgrounds to pursue careers in computer science. Perhaps surprisingly, not all role models are created equal.[46]

Mentors often have a prior connection with students, helping them see what CS looks like within their own communities. My work has led me to believe that mentors are most effective in the early years (K–5), when most students still believe that anything is possible. Mentors, especially in the form of parents, tend to lose effectiveness around middle school, when increased autonomy causes students to start moving in the opposite direction from where they are guided. Fortunately, mentor relationships tend to come back into favor once students leave home for college.

Role models, on the other hand, are generally "special guests" who are held in esteem by students or the community in which the students live. Generally, role models are introduced through stories, videos, online chats, or even classroom visits. Students don't necessarily need to meet role models to aspire to be like them, so it's perfectly fine to pick out examples from history, like Ada Lovelace (Figure 7.2), Grace Hopper (Figure 7.3), Annie Easley (Figure 7.4), or Hedy Lamarr (Figure 7.5).

Figure 7.2 Ada Lovelace

Source: Etching by William Henry Mote, 1838.

Figure 7.3 Grace Hopper

Source: United States Navy.

Figure 7.4 Annie Easley

Source: NASA.

Figure 7.5 Hedy Lamarr

Source: Focal Press (2001), p. 211.

Unfortunately, renowned role models will only get students so far. They tend to get raised to an elevated standing and are presented as geniuses with few flaws. At some point, students may stop identifying with those heroes, no matter how similar they are in demographic, because their status feels unobtainable. This is where community role models become helpful. Allowing students to engage with local adults or near-peers can bring a level of reality to a CS career that students may not easily find elsewhere.

Promotion and Advertising

Students can't take a class if they don't know it exists—and they won't want to take a class they think their friends won't want to take. You can greatly increase the odds of appealing to a diverse selection of students by hyping your class all year long.

▶ **Create a Traveling Show:** Consider promoting your class by showing off projects with students in other classes. Try hosting a CS carnival for your grade or posting portfolio images in the halls. This develops a sense of anticipation for other students who then look forward to the day they can take your class, too.

▶ **Graduation Awards:** If your school announces graduation awards or has student ceremonies, see if you can get together with other teachers who offer computer science to give a CS distinction. Reward students for creating community games or solving local problems using computer science. This helps bring the subject out of the shadows so others can see that the community values the skill.

▶ **Communicate With the Guidance Counselor:** Guidance counselors often have the faulty assumption that computer science is only for certain kids. Have a one-on-one with your school guidance counselor to make sure that they're pointing students your way. It might be worth going one step further and suggesting that the guidance counselor familiarize themselves with all of the various ways that computer science can enhance a student's life, whether or not they intend to have a tech career.

> **Peer Champions:** Recruit peer champions. You may be able to find previous students who are willing to post about their time in your class, or you can seek out potential students and ask them to help you recruit friends to take a future class. When you do, make sure to tap into a variety of sources so that your champions reflect the makeup of the class you strive to teach.

Balance Your Unconscious Biases

There are also a bunch of classroom tricks you can use to help students feel welcome on the daily.

> **Diversify Your Teaching Style:** Vary the medium that you use when presenting material to your class. Some students feel more comfortable with video, some with worksheets, and some with project-based activities. When possible, try to provide multiple options for graded work (oral presentation *or* written report *or* poster project). Computer science offers options for students with visual impairments and hearing impairments, as well as those with dyspraxia, or moving impairments, so don't underestimate what students can accomplish when you match activities to their strengths.

> **Randomize Selections:** When calling on students freeform, unconscious biases can affect who you select and how often. Curb the urge to avoid introverted students or lean on extraverted students by using randomized selection methods like drawing names from a cup, using random-number websites, or moving through students in order, desk by desk.

> **Diversify Classroom Examples:** Avoid using exclusively sports metaphors or always gendering the subjects of your examples in a binary way. Try pulling metaphors from all walks of life and using they/them pronouns when possible.

> **Pay Attention to Compliments:** If you frequently pay compliments, make sure you pay the same types of compliments across the board. When you praise, you should equally praise all students for the same skills consistently. One way to do this is to have "compliment category days." For example: Monday is always creativity, Tuesday is always perseverance, etc. This ensures that you don't highlight how hard an athlete tries while highlighting how pretty a singer's final product is.

> **Call Out Prejudiced Behavior:** When you hear students express prejudicial ideas, like boys suggesting that girls be in charge of the artwork on a CS project while boys handle the coding, try gently redirecting with more equitable suggestions. If the scenarios persist, consider addressing the ideologies behind them directly to the entire class.

> **Do the Work:** In the early stages of allyship, we don't know what we don't know. The quickest way to attune yourself to the microaggressions happening in your community, your school, and your classroom is to research psychological safety, social injustice, and code switching.

Play Time

Fun With Equity!

Adding equity can be fun! If you've never visited WheelOfNames.com, take a second to hop over there and give it a try. It's an excellent way to make sure that students are being called on fairly, and you have the opportunity to adjust the list of names at any time.

Occasionally, resources become unavailable. If WheelOfNames.com is no longer operational, try searching for "wheel of names" on the internet. You may find a functional substitution, like PickerWheel.com or SpinTheWheel.io.

Source: OmanderConsulting LLC.

A Word on Equity Compared With Justice

Recently, I had the opportunity to speak with Shana White, the Director of CS Equity Initiatives for the Kapor Center. During our conversation, she spoke about the role of **justice** within computer science.

For many years, educators in this space have been working to promote equity over equality to help people understand that it's not enough simply to give every student identical support and opportunity. To truly strengthen our communities, we need to give more to those who have been identified as underserved. By focusing more resources on the underserved, we even the playing field and create a landscape that's more fair and balanced.

Over the last couple of years, focus has begun to shift toward justice over equity. This is to acknowledge that the reason we've been forced to distribute resources in this way is that the underlying system is unfair. If we're going to focus on fixing the diversity and inclusion issues within CS, then we need to work toward solving the root causes of inequity and not just mediate the symptoms of it.

This idea of justice in CS extends into **social justice**, which is by definition the implementation of a fair and equal society where every person is valued and political and economic decisions are balanced and honest. Translating that into the world of computer science means that schools would remove any and all barriers to entry or success and tech companies would be proactive in eliminating biases in algorithms, hiring practices, and workplace culture.

As an added bonus, the word *justice* was intrinsic to the foundation of the United States of America, meaning that the word would be very difficult to ban by politicians or administrators who object to the idea that students should be taught about differences experienced by people of various colors or religions.

Many of us have seen an illustration similar to Figure 7.6. The original has always been problematic for me since it appears to depict people trying to watch a baseball game without a ticket. Here, I've created one that's less questionable, and put it in the public domain so that it can be used in slide decks and presentations. You can find the full-resolution, full-color version on my blog at https://medium.com/geek-groupies.

https://medium
.com/geek-groupies

Summary

This chapter was packed with information and may benefit from a second readthrough later in your planning process. Whatever your familiarity with inclusive practices, remember that equity, inclusion, diversity, and justice are vital components of computer science whether the subject stands alone or is integrated into another class. Today's students will be tomorrow's educators, innovators, and lawmakers. They need to understand how algorithms and technological policies affect the communities in which they live.

Diversity is not a token gesture within computer science—it's foundational to innovation, ethics, and social responsibility. We have already stumbled into real-world examples of biases in technology, some of which are

Equity, inclusion, diversity, and justice are vital components of computer science whether the subject stands alone or is integrated into another class.

Figure 7.6 Reality, Equality, Equity, and Justice

Source: Kiki Prottsman, 2024. Created using Photoshop with elements from Midjourney and edited with graphic design programs.

mildly irritating while others could mean the difference between life and death. These biases often stem from a lack of diversity among development teams.

Beyond the prevention of biased algorithms, diverse teams lead to better problem-solving and increased profitability. Interestingly, the impetus for these team demographics is shaped at the K–12 level, as students often decide what they believe they're good at quite early in life.

Reflection Questions

1. As a teacher who plans to integrate computer science into a different specialty, you may find yourself in a place where students are already lining up to take your classes, which is a fantastic way to give CS a positive reputation in your school. Even so, did you find any techniques in this chapter that you can utilize to make your current classroom more welcoming and inclusive?

    _____

    _____

    _____

    _____

2. Looking at the stories at the beginning of this chapter that highlight ways that technology has failed large swaths of the population, can you think of any projects you could assign students that would inspire them to come up with a proposal for a solution? Even without the addition of any hardware or programming, such a project would likely meet CSTA Standard 2-IC-20 (Compare tradeoffs associated with computing technologies that affect people's everyday activities and career options).

_____

_____

_____

_____

3. If you have a dedicated STEM or CS teacher in your school, consider discussing with them ways you can help inspire students to sign up for their class. This could include inviting students to your room for a show-and-tell, partnering with the instructor on a cross-curricular project, or championing the CS class to your students as a way of rounding out their skill sets. To whom do you need to reach out? What would you like to propose?

_____

_____

_____

_____

Always Be Evolving: ChatGPT and AI

CHAPTER #8

Q: How do I tell if my students have been using ChatGPT to help with their assignments?

A: I'd like to dig deeper here and ask if that's really the problem we should be worried about. Keep in mind that ChatGPT is not a scholar. It's not equivalent to having someone with a PhD complete assignments on behalf of your students. For students to turn in a convincing paper or correctly solve a problem using ChatGPT, they'll need to know how to properly formulate their prompts, review answers for inaccuracies, change the tone of the writing to match their own, and double-check all of the references to be sure the system didn't outright make them up. These are all valuable skills, especially for the world into which your students will be growing.

Instead, consider allowing students to use ChatGPT responsibly by citing the tool and showing their work. Giving students an honest path to follow will help you understand what level of assistance they felt they needed and how much energy they put into understanding the problem at hand.

> *Consider allowing students to use ChatGPT responsibly by citing the tool and showing their work.*

What Is Artificial Intelligence— and What Isn't It?

One of the unique and amazing things about computer science is the way it's always changing. This can get awkward when our administration wants us to teach the exact same things in the exact same way year after year.

My challenge to you is to keep your eyes open for new technologies and new trends. Then, instead of fretting over all the ways that the technologies can change the landscape, try to understand the skills that students will need to use the new technologies responsibly. We can't keep tech from moving forward, but together we *can* shape what the future of tech looks like. This is especially true for the burgeoning subject of artificial intelligence (AI).

As much as I try to avoid these in real life, I'd like to interject here with a "well, actually ... ".

While I normally find comments like these pretentious and unnecessary, some things benefit from clarification—artificial intelligence is one of those things. So, here it goes:

Well, actually … ChatGPT isn't AI, exactly.

ChatGPT isn't AI in the way we've been trained to fear it. ChatGPT (along with its friends **Dall-E**, **Midjourney**, and Bard) are in a category called large language models, or LLMs. These tools have been nicknamed "generative AI," because they synthesize large amounts of data into new and interesting pieces, which gives the illusion that they're creating works on their own. It's the same technology that's been used by chatbots since the late 1960s. While algorithms have become stronger and faster over the years, they are still a long way from being able to take over the world.

For something to truly be artificially intelligent—we're talking **HAL9000** or "**three laws of robotics**" level—algorithms need to be capable of learning autonomously, working without human intervention, and making decisions on their own. Google flirted with that ability in 2020 when they created a robot that successfully taught itself to walk,[47] but that is not the technology being used in the latest wave of generative models.

Generative AI is "trained" on large data sets, then provided with parameters that can be used to reformulate sample data into something new. ChatGPT-3, for example, has roughly 175 billion parameters governing its algorithms;[48] ChatGPT-4 is thought to have nearly 1,000 times as many. This means that humans feed it data, humans tell it what it can and can't do, humans ask it questions, and humans tell it whether the results were good or bad. There is very little about an LLM that is autonomous. Instead, you can think of an LLM as a robust search engine that sometimes lies to you and gives you attitude.

This isn't to say that LLMs are useless. On the contrary! The ability to pull apart and resynthesize data is extremely beneficial, especially when it comes to idea generation or comparing a piece of work to previous exemplars.

That said, I'll continue to refer to LLMs as AI throughout the rest of this chapter, since that's how they're currently being billed in the wild.

Spotlight

From Kiki's Life

Some time ago, I decided to use ChatGPT for a rather unconventional purpose: matchmaking advice. That's right, I was looking for love—or, at least, a decent date for Saturday night. So, I opened the ChatGPT interface and typed, "How do I impress someone on a first date?"

ChatGPT responded with a list of tips, ranging from "Be yourself" to "Discuss interesting topics" and even "Wear something that makes you feel confident." Feeling optimistic, I followed the advice to the letter and prepared for my date.

The date started off well. We met at a cozy café, and the conversation flowed naturally. I was feeling good until my date asked, "So, do you have any interesting hobbies?" Remembering ChatGPT's advice to discuss interesting topics, I enthusiastically blurted out, "I love conversing with AI models and exploring the ethical implications of machine learning!"

To my surprise, my date's eyes lit up. "No way, I'm a software engineer working on AI ethics!" We spent the rest of the evening engrossed in a passionate discussion about AI, data privacy, and even ChatGPT.

Feeling like I'd hit the jackpot, I decided to push my luck. I pulled out my phone and said, "Would you like to ask ChatGPT a question? It's how I prepared for our date!"

My date chuckled and typed, "ChatGPT, how did I do on my first date?" The AI responded, "I can't observe human interactions, but if you're asking me, it must have gone well!"

Here's where it gets wild: My date revealed that they were actually part of the team that developed ChatGPT! Not only that, but they were so amused by our interaction with the AI that they invited me to give a talk about "Finding Love in the Age of AI" at an upcoming tech conference.

The talk was a hit, and it led to a collaboration with ChatGPT's development team on a new feature: AI-assisted relationship advice. And as for my date and me? Let's just say we're exploring the algorithm of love, one ChatGPT conversation at a time.

Educator Takeaway: The lesson here? Sometimes, AI can play Cupid in the most unexpected and hilarious ways, leading to opportunities—and love connections—you never saw coming.

If you read the "From Kiki's Life" segments, you may be wondering why I would make such an odd choice as to slip a personal story about my love life into a book on integrating computer science into the classroom. Well, I didn't: ChatGPT did!

To illustrate the power of using an LLM for idea generation, I decided to give ChatGPT a prompt, then include the unedited result. The final story, reproduced in the previous Spotlight feature, was fairly well-written and believable, so let me tell you how I set it up.

First, I fed ChatGPT a sample of my writing from earlier in this book. Then, I crafted a prompt that would use that example to shape a new story:

> Please use this sample to identify my tone, then write a piece that describes an interesting and funny interaction with a Large Language Model (like ChatGPT, Dall-E, Midjourney, or something completely new). I would like you to include an interesting story, some exposition about what my thoughts were during the interaction, a funny result, and a lesson that I learned about using AI due to the interaction.

Keep in mind, writing a good AI prompt is a lot like programming. Garbage in, garbage out. The more time and description you put into your initial request, the better chance you have of getting an authentic and entertaining final product.

After this test, I asked it to generate additional options. Some were on the boring side (one where I asked Dall-E for a couple of crazy pictures and it complied) and some were downright unbelievable (Midjourney started predicting the future and we broke into alternate universes).

Since my story was fiction, I didn't need to validate any points for truthiness—but had this been for a research paper, I likely would have used AI to help with that chore, as well. In fact, Bing now has a helpful feature where you can chat with an LLM and ask it to find papers validating points or counterpoints. **But be careful!** Many of these models were trained using posts from Reddit, so it's worth checking that the cited papers are real and from a trusted source.

Finally, I ran my story through a plagiarism checker because I would hate to unknowingly publish text that was pulled from another creator. My piece was evaluated as 100% unique, and since OpenAI statutes allow for publication of responses, I was set to move forward with including my cute little tale.

I hope you can see how many lessons were embedded in the exercise described above. Let me pull out a handful:

▶ Crafting a good prompt for AI is challenging and important

▶ AI can give you multiple results for the same prompt

▶ You can't believe everything/anything AI tells you (AI makes things up and lies)

▶ AI output is not always original or unique

Combine those with the knowledge that LLMs are trained on data that is implicitly biased, and you have inspiration for a bunch of cutting-edge exercises that can be woven into your everyday activities to help educate students on the proper care and feeding of AI.

Play Time

OpenAI is responsible for ChatGPT and Dall-E. Both are extremely useful for educators.

If you haven't played with generative AI yet, you're only hurting yourself at this point. There is so much busywork that it's able to do for you. From writing the first draft of a difficult email to creating images for your slides, there's no reason to toil away in the same way that you used to.

Head over to OpenAI.com and click "Get started" in the upper-right corner to make a free account (you can use a **burner email** if you like, but it will require a phone number and each number can only be linked to one account).

Once you have your account, it will ask you which product you want to explore. Choose from ChatGPT or Dall-E. Whichever you choose, start with a simple prompt, then make it more and more complex and see what happens.

Dall-E Example:

Prompt #1: I'd like an image of a watermelon, please.

Prompt #2: I'd like an image of a perfectly round watermelon, please.

Prompt #3: I'd like an image of a perfectly round watermelon sitting in a field of sunflowers, please.

Prompt #4: I'd like a photo-real image of a perfectly round watermelon sitting in a field of sunflowers, please.

Prompt #5: I'd like a photo-real image of a perfectly round watermelon sitting in a field of sunflowers with a sunset in the distance, please.

Note: At the time of this writing, you are gifted 15 free Dall-E credits each month. After that, you have the option to purchase 115 credits for $15 USD. See what I created here:

Source: Kiki Prottsman, 2024. Created using Dall-E3.

Integrating LLMs Into Classroom Assignments

I'm a huge fan of being open with students about the pitfalls of AI. It helps students think more critically about the way they use the tools, and it inspires them to do more than regurgitate whatever ChatGPT replied to their prompt. Sure, it might be harder to catch students using LLMs once they know that text needs to be double-checked and polished, but the finishing work is what ultimately allows the students to become familiar with the core material.

If you're interested in steering students toward the most acceptable ways to incorporate LLMs into their assignments, consider *requiring* their use on certain activities when brainstorming. Here are some examples of exercises that promote the use of LLMs.

Geography: Learning About Captain James Cook

American schools have a game called *The Oregon Trail* to help teach them about the famous expedition by Lewis and Clark. You've been asked to create something similar for explorer Captain James Cook.

1. Read a short biography of Captain James Cook

2. Using a worksheet, design three separate prompts that ask ChatGPT to come up with an idea either for a board game or a video game based on the life of the explorer. Make sure the prompts include:

   a. The game you want used as an example

   b. The description of the explorer that you want the game to consider

   c. The purpose of the game (teach the history of the voyages of Captain James Cook)

   d. The audience you want the game to be created for (Elementary? Middle school?)

3. Select one of your prompts to enter into ChatGPT and record your favorite response from the tool

4. Choose one element from the suggested game to bring to life for the rest of the class; this could mean:

   a. *Video game:* A live-action simulation of what it would look like to watch the game being played

   b. *Video game:* A poster showing a screenshot from one of the in-game activities

   c. *Video game:* A slide show that illustrates the voyage that the player will take as they play

    d. *Board game:* A sample game board that shows interesting points of action along the way

    e. *Board game:* Game cards that challenge players with trivia around Captain Cook's interactions

    f. *Either:* A commercial or meme marketing the finished game

5. Follow the rubric to make sure you have all the features required in this assignment, including:

    a. Names of other people, places, and ships involved in the adventure

    b. At least five facts commonly highlighted around the life and times of Captain James Cook

    c. Visuals that include accurate representation of some location, foliage, or wardrobe included in the journey

6. Verify all of your facts using at least three resources! Remember that ChatGPT is often wrong!

Grammar, Tone, and Writing

This assignment will help you rework a piece of writing into something suitable for your portfolio.

1. Fill in the blanks in this paragraph to create a prompt for ChatGPT on the famous person of your choice.

   "Please use Old English to write a 300-word story about _____ (a person). Include at least four facts about their life, including their date of birth and the highlight of their career."

2. Print the resulting story on paper.

3. Circle all of the provided facts, then look each of them up. Place a checkmark by any fact that can be verified by two or more sources. Place an X by any facts that turn out to be false.

4. After removing or correcting all bad information, translate the remaining work using your own words into the common vernacular of the day (or proper English, or Spanish, or whatever suits your class).

Art History

1. Select one of the paintings that we've been studying this term and ask Midjourney to reimagine it in another artist's style (example: Figure 8.1, *Mona Lisa* in the style of Edvard Munch).

2. Print one of the resulting images and grade the blend (see Figure 8.1). What was done well? What could have been different?

Figure 8.1   Using Dall-E3 to Blend the styles of Leonardo da Vinci and Edvard Munch.

Source: Kiki Prottsman, 2024. Created using Dall-E3.

3. Recreate (or improve) the blend on your own, using paint, pencils, chalk, or pastels.

4. What three things evoked feelings of the original painting?

5. What three things allowed the style of the new artist to come through?

Computer Science

It won't be long before generative AI begins helping students learn to code.

Right now, the emphasis in this space often lands on the potential for students to use AI to cheat on their assignments, but I believe we'll soon move to a model where classrooms rely on generative AI to introduce computer science at the most basic level. Students can use AI to generate multiple solutions, then describe their process for determining which solution to use. There's validity in learning to analyze existing code or hardware designs before learning to create original solutions.

Much like block-based programming, generative AI can be embraced as a level-appropriate tool for helping students to understand the basics before they've gained the skills they need to dig in more deeply. Personally,

I'm excited by this prospect because I believe that ideas really have the potential to click when people start explaining why they prefer one method over another.

AI and LLMs to Assist Student Learning

Beyond stand-alone AI sites, LLMs are being used in a new batch of tools created to help remediate and assist students with critical skills.

In 2023, Microsoft released a suite of digital learning tools that aimed to strengthen reading, writing, math, and communication. These applications, called "Learning Accelerators," have been embedded into Microsoft 365 products, allowing students to overcome discomforts and disabilities using Immersive Reader, Dictation features, and even a Math Assistant that walks through equations to help students understand the steps they need to take to solve a problem.

Around the same time, Google embedded Duet AI into Google Workspace products, highlighting features that help students get past the "blank page" stage. These tools can generate first drafts of letters, create sample spreadsheets, or develop unique slideshow images from text prompts.

Similar technology has been embedded into coding platforms, including Ghostwriter from Repl.it and Copilot from Visual Studio Code. Both go beyond traditional developer environments to make suggestions not only around what it deduces you should type to complete a line, but what it deduces you may want to type to complete a feature. These predictions are based on thousands of previous programs that shared similarities with your current project.

Tools like these are meant to keep students in a state of flow, helping them stay engaged in the work they're doing and participate in productive struggle, instead of getting lost in a place where they have no idea what to do next. But, before students can benefit from these tools, someone needs to show them how they can be used responsibly.

> *"When the winds of change blow, some people build walls and others build windmills."*
> *— Ancient Chinese proverb*

AI and LLMs to Assist Teachers

Let's jump into an area where AI really shines: saving teachers time!

There are so many tools for teachers nowadays—tools that help you grade writing; tools that help you create lesson plans, quizzes, worksheets, or slides. There are even tools that can record your lectures for you so students can search through the transcripts for answers to their own questions. Here are a few that I think have the most potential.

> ❯ **Consensus.app**
>   Think of this as a search engine for facts. It helps you find research-based answers to important questions, complete with reference papers.

▶ **Curipod**
Curipod allows teachers to easily create slide decks for lessons that include questions, word clouds, and polls.

▶ **Education Copilot**
As the name suggests, this tool is meant to help teachers with work they would otherwise need to do themselves. Claiming to help streamline educator planning and prep, Education Copilot offers templates for lesson plans, handouts, project outlines, writing prompts, student reports, and more.

▶ **Fetchy**
An AI-assisted tool for creating lesson plans, frameworks, permission slips, comprehension questions, newsletters, and more. It claims to make teachers feel like they have access to a personal assistant.

▶ **Formative**
This app helps to autogenerate and grade assessments, providing real-time feedback from student work and allowing for targeted interventions.

▶ **MagicSchool.ai**
Similar to Fetchy, MagicSchool.ai provides a suite of teacher assistant tools. This application boasts a track for special education teachers with IEP paperwork and differentiation. It also allows for generation of rubrics, a text-leveler to make sure your assignments are age-appropriate, and a tool that helps pull keywords for definitions and pre-teaching.

▶ **Otter.ai**
Otter.ai transcribes live or virtual lectures so students can search for information while completing assignments. Notes can be highlighted and images can be added to illustrate points. It even automatically generates a summary.

▶ **Yippity**
The Yippity site lets you automatically generate quiz questions or flashcards from a block of text or a website.

Beyond all these apps, ChatGPT can also be used for quite a lot, especially with ChatGPT-4. It's now possible to feed the model a bunch of information and have it suggest ideas for lessons, worksheets, quizzes, and flashcards.

Do keep in mind, AI still has a long way to go and it's extremely fallible. In the same way your students need to pay attention to the results that they receive, you'll need to do a quick gut check on your materials. And for goodness' sake, do *not* ask ChatGPT if it wrote your student's papers. It will often say "yes," even when the content came directly from the student's mind, and *you* may end up needing to report to the principal's office![49]

Addressing Hopes and Fears

Early in 2023, I went to a conference where a panel of businesspeople asked the audience for their "hopes and fears" around AI. Being that we were at the beginning of the LLM onslaught, the educators had a whole lotta fears and were light on hopes. Unfortunately, the panel glossed over the teachers' concerns and dove into a spiel about how AI was the way of the future. In my opinion, this was a huge miss on their part. Teachers are not going to welcome something with open arms when they think it's going to be harmful to their students.

For nearly as long as artificial intelligence has been public knowledge, it has been villainized. From Ray Bradbury's short story "The Veldt" in the 1950s to the films *2001: A Space Odyssey*, *War Games*, and *iRobot*, we've been primed to worry about what technology is capable of, the ultimate fear being "**the singularity**," where technology becomes sentient and no longer controllable by humans. While we're nowhere close to that point, all this talk about AI is giving people premonitions—in large part due to the possibility of technological autonomy. Fortunately, we should be able to use the lessons we're learning with generative AI to help regulate the real thing when it comes along.

No matter what the opinion of the general public, LLMs will continue to get more robust and convincing. It shouldn't be any surprise that they are being hailed as the biggest advancement since the internet—and the internet changed *everything*. In the same way, we can pretty much guarantee that AI will be taken advantage of, misused, and abused. But that doesn't mean we need to encourage people to avoid it. It means we need to train people to use it responsibly.

The last decade has been full of discussions around AI safety and precautions. Part of this has been driven by the desire to ensure the safety of AI systems and mitigation of potential risks. The most recent regulations have been aimed at protecting creators of sourced content and copyright issues. This work led to a document of expectations, created by the White House in July 2023, that lays out a plan for "Safe, Secure, and Trustworthy AI." It also inspired the first annual AI Safety Summit at Bletchley Park in the UK in November of 2023.

While this might help a bit with the underlying fear of world domination, we still have a slew of other concerns to address. That list includes the worry that students will use AI to cheat, the very real possibility that AI will provide incorrect answers to students who are doing research, and the potential for copyrighted material to be passed along with the assumption that it can be used in the work of others.

Cheating With AI

Right now, it's extremely easy to cheat with AI because the idea of cheating isn't well defined in this space. Big-name corporations are asking employees to utilize AI to kickstart letters, emails, and documentation wherever it makes sense—and at the same time, many teachers want students to avoid getting *any* help from AI. Moving forward, it's going to be increasingly difficult to keep AI out of the educational process, so my advice is to focus on responsible ways to use the technology that minimize tedious busywork while still allowing students an opportunity to be creative and showcase the knowledge that's being assessed.

This area should continue to mature rapidly, so I fully expect a wave of articles, lesson plans, and best practices around the subject in the months to come.

Deceptive AI

The current level of trustworthiness in AI-generated responses is quite poor. The internet is full of stories of lawyers who were given false briefs, reporters who were goaded into doubting their partners, and teachers who were incorrectly told that all the papers submitted for term finals were generated through ChatGPT.

AI lies. AI makes things up. AI interprets facts incorrectly. Whether this trend will get better or worse is yet to be decided, but the fact remains that we should be very skeptical of anything that an LLM passes along as a fact.

Students should be taught to verify "factual" points in multiple ways before incorporating the ideas into their assignments. This is a great place for a lesson on trusted resources versus untrustworthy sources of information. This might also help fortify students against false ideas as they get passed around social media.

Utilizing Copyrighted Materials

At the time of this writing, ChatGPT, Dall-E, and Midjourney all have policies allowing users to incorporate the results of prompts into future works, but they do not guarantee uniqueness of their products. For that reason, it's important to run a search on any content you receive before you incorporate it into work of your own.

There are several antiplagiarism sites that can be used to check text provided from ChatGPT, but verifying the individuality of an image is a bit more difficult. However, since the source images fed into products like Dall-E or Midjourney are digital, nearly all of them should exist on the web. This means that you can use a tool like TinEye that crawls the web to try to find something similar. You can also go to google.com and click the camera icon in the search field to upload your picture and "search by image."

While it's very likely that neither of these methods will find anything identical to what you're trying to verify, images don't need to be identical to infringe on copyright. The laws vary around the world, so make sure you can't find anything "reasonably similar" to the image you plan on using. If all of these checks come back in your favor, then it's very likely you can move forward with caution.

Summary

Technology is still evolving, and it's far from perfect. As educators, we have the opportunity to help shape the way our community uses technology. We also have the opportunity to help our community become a part of the technology creation process, which is the only way we can ensure that the biases, ethics, and functionalities behind the products work in our favor.

Ushering students through the 21st century is not a one-person job. Teachers need support from parents, administrators, and fellow educators. That's why it's so important that we reinforce the importance of computer science from all angles and in all kinds of classrooms, because the people who are needed most in this industry are also the people who are least likely to feel welcomed and included in the space if they arrive on their own later in life.

Even for the students who have no intention of working in the computer science industry, a firm foundation in CS will help them develop skills that have been identified in successful people: communication, persistence, problem-solving, creativity, and collaboration.

Let's work together to change the landscape of technology—and in the process, craft some fun, educational, and unforgettable lesson plans.

Reflection Questions

1. How much experience do you have with AI and LLMs? List five of these tools below, and if they don't easily come to mind, head to the internet to look some up.

   _____

   _____

   _____

   _____

2. Choose one of the LLMs from the previous question that you are least familiar with. Write down an idea for how you could responsibly use it in your classroom to help you with a lesson.

   _____

   _____

   _____

   _____

3. Choose one of the LLMs from the first list and write down an idea for how you can responsibly use it with students in a fun assignment that fits in with your core content.

_____

_____

_____

_____

Afterword

I had such a great time writing this book for you, and I wish I could be there to see you put into practice all the things you've learned. Nevertheless, I hope you send me back some love by posting your successes (and failures!) on social media and tagging @KIKIvsIT. We're all in this together, across time and space.

https://medium
.com/geek-groupies

I covered a lot inside these pages, but there was so much more that didn't make it to print. If you're interested in learning more, or just want to see where my research spirals took me, subscribe to my blog at https://medium.com/geek-groupies to read additional stories and see some of my outtakes.

I'd also love to stay in touch with you as I launch new projects. If you're interested in being the first to know when new releases are coming up, please hop over to KIKIvsIT.com and sign up for my newsletter. It's a great way to begin a two-way conversation so that you can keep me up to date with your questions and comments, as well.

I hope to see you at a conference soon!

Cheers and all the best,

Kiki

Glossary

· ·

abstraction: The process of simplifying complex systems by focusing on essential features and ignoring irrelevant details.

active learning: A teaching method where students engage in activities like discussions and problem-solving to learn, fostering critical thinking, and practical application of knowledge.

AI: The simulation of human intelligence in machines, enabling them to perform tasks that require decision-making.

algorithms: Step-by-step procedures or formulas for solving problems.

backward design: A planning approach where you start with the end goal and work backward to develop the curriculum or program.

binary: A structure with only two options (generally some form of true and false, off and on, or 1 and 0).

block-based: A type of programming where code is represented as visual blocks that can be dragged and dropped.

burner email: A temporary email address used to avoid the government, spam, or people you want to communicate with once and then never again.

char: Short for "character," a data type that stores a single symbol in programming languages like C and Java.

ChatGPT: A conversational AI model trained to generate human-like text (and now images, tables, sheet music, and more) based on the input it receives.

coding: The act of writing programs for entities in a language that the entity best understands.

cognitive load: The mental effort required to learn or perform a task.

computational thinking: A problem-solving method that involves breaking down complex problems and using algorithms to solve them, as if you were trying to format the problem to be solved by a computer.

computer science: The study of computers and computational systems, including software, hardware, and algorithms.

conditionals: Statements that help a computer "decide" whether to execute code based on if a condition is true or false.

constructionism: A learning theory that suggests children learn more effectively when they construct something physical.

constructionist: An advocate or follower of the constructionism learning theory.

Dall-E: An AI model trained to generate images based on textual descriptions.

data structures: Ways to organize and store data in the memory of a computer.

decomposition: Breaking down a complex problem or system into smaller, more manageable parts.

digital: Pertaining to data stored and processed in a binary manner.

EdTech: Short for Educational Technology, the use of technology to enhance learning and teaching.

emotional literacy: Understanding and managing one's own emotions enough to know when it's time to take a nap.

equality: Ensuring everyone has the same resources and opportunities.

equity: Providing individuals with the resources and opportunities they need to reach an equal outcome.

Escalator Method: Approaching students where they are educationally and introducing new topics only when the previous topic has been connected to other elements of their lives.

fiero: A term for the feeling of triumph over adversity.

foundational computer science: Enough experience with computer science to include the understanding of basic definitions and concepts.

frustration: The feeling of being stuck or unable to solve a problem, often accompanied by additional emotions like anger, embarrassment, or physical reactions.

generative AI: AI models that can generate new data, such as text or images, based on what they've learned.

grok: To deeply understand a concept, system, or technology to the point where it becomes second nature (originating from the science fiction novel *Stranger in a Strange Land*, by Robert A. Heinlein).

growth mindset: The belief that instead of being "good" or "bad" at something, abilities and intelligence can be developed over time through practice.

HAL9000: A fictional AI character from the movie *2001: A Space Odyssey*, known for its advanced capabilities and ethical dilemmas.

hardware: The physical components of a computer system, like the hard drive, memory, and keyboard.

implicit bias: Unconscious attitudes or stereotypes that affect decisions and actions (another term for *unconscious bias*).

impostor syndrome: The feeling that one's achievements are undeserved and the fear of being exposed as a "fraud," even when quite skilled.

Inquiry-Based Learning: A teaching approach where students learn by asking questions, investigating, and exploring topics, leading to a deeper understanding through active discovery and problem-solving.

justice: Fair treatment and equitable distribution of resources and opportunities, often discussed in the context of technology's societal impact.

LLMs (Large Language Models): Advanced AI models trained on vast datasets that are designed to generate new creations based on previous "learnings."

loops: Programming constructs that repeat a block of code until a certain condition is met.

Midjourney: An advanced AI model trained to generate images based on textual descriptions and links.

pattern matching: Identifying sequences or regularities in data.

productive struggle: Having a difficult time with a task, but continuing to make progress nonetheless.

programming: The process of creating software by conceiving of and writing code.

social justice: The pursuit of equality and fairness as a key feature in society.

software: Programs and applications that run on computer hardware.

STEM: An educational approach focused on Science, Technology, Engineering, and Mathematics.

STEAM: An educational approach that integrates Science, Technology, Engineering, *Arts*, and Mathematics.

STREAM: An educational approach that integrates Science, Technology, *Reading*, Engineering, Arts, and Mathematics.

STREAMERS: The acronym that we're sure to end up with if we continue being required to justify the importance of non-siloed educational subjects in K–12 education.

Teachers Pay Teachers: An online marketplace where educators can buy and sell educational resources.

technology: The application of scientific knowledge for practical purposes.

text-based: Programming or interaction that involves typing text, as opposed to using graphical elements.

"the singularity": A hypothetical point in the future when technological growth becomes uncontrollable, often associated with AI surpassing human intelligence.

"three laws of robotics": A set of ethical guidelines for AI and robots, introduced by science fiction writer Isaac Asimov. In paraphrase: A robot can't hurt a human or allow a human to get hurt by abstaining; a robot must listen to humans unless it would break the first rule; a robot must keep itself safe, but not if it breaks the first or second rule.

unconscious bias: Attitudes or stereotypes that one is unaware of, though they manage to affect one's decisions and actions (another term for *implicit bias*).

unplugged: Activities that teach computational thinking or computer science concepts without the use of a computer.

Bibliography

1. Bureau of Labor Statistics. May 2022 National Occupational Employment and Wage Estimates. Retrieved 22 October, 2023, from bls.gov

2. Arfé, B., Vardanega, T., Montuori, C., and Lavanga, M. "Coding in primary grades boosts children's executive functions." *Frontiers in Psychology* 10 (2019): 2713. https://doi.org/10.3389/fpsyg.2019.02713. PMID: 31920786; PMCID: PMC6917597.

3. Century, J., Ferris, K. A., and Zuo, H. "Finding time for computer science in the elementary school day: A quasi-experimental study of a transdisciplinary problem-based learning approach." *International Journal of STEM Education* 7, 20 (2020). https://doi.org/10.1186/s40594-020-00218-3

4. Camp, T., and Gürer, D. "Investigating the incredible shrinking pipeline for women in computer science." ACM Committee on Women in Computing (1997).

5. Master, A., Meltzoff, A. N., and Cheryan, S. "Gender stereotypes about interests start early and cause gender disparities in computer science and engineering." *Proceedings of the National Academy of Sciences* 118.48 (2021): e2100030118.

6. FitzGerald, C. et al. "Interventions designed to reduce implicit prejudices and implicit stereotypes in real world contexts: A systematic review." *BMC Psychology* 7.1 (2019): 1–12.

7. Dee, T., and Gershenson, S. "Unconscious bias in the classroom: Evidence and opportunities." Google's Computer Science Education Research (2017).

8. Skaalvik, Einar M., and Sidsel Skaalvik. "Self-concept and self-efficacy in mathematics: Relation with mathematics motivation and achievement." *The Concept of Self in Education, Family and Sports* (2006): 51–74.

9. Education worldwide—statistics & facts. Statista. Accessed 23 October, 2023. https://www.statista.com/topics/7785/education-worldwide/

10. BBC News. "What hours do teachers really work?" BBC. https://www.bbc.com/news/education-27087942. Published 19 April, 2014. Accessed 23 October, 2023.

11. The Association for Women in Science. "Mentoring means future scientists." Washington, D.C., July 1993. http://www.awis.org

12. Prottsman, C. L. L. "Computational thinking and women in computer science." Master's thesis, University of Oregon. Retrieved 16 April, 2023, from https://www.learntechlib.org/p/117630/

13. Lazzaro, N. "Why we play: Affect and the fun of games—Designing emotions for games, entertainment interfaces, and interactive products" (2012).

14. Liu, Z. et al. "Sequences of frustration and confusion, and learning." *Educational Data Mining* 2013 (2013).

15. McGonigal, J. Reality Is Broken: Why Games Make Us Better and How They Can Change the World. London: Jonathan Cape (2011).

16. Liquin, E. G., and Gopnik, A. "Children are more exploratory and learn more than adults in an approach-avoid task." *Cognition* 218 (2022): 104940.

17. Hoff, M. "Here are the best high-paying and fast-growing jobs for the next decade." *Business Insider*. Published online 12 September, 2023. Accessed 23 October, 2023. https://www.businessinsider.com/best-high-paying-fast-growing-jobs-careers

18. Bls.gov. Accessed 23 October, 2023. https://www.bls.gov/emp/tables/stem-employment.html

19. U.S. Census Bureau. "STEM majors earned more than other STEM workers." Published online 2021. Accessed 23 October, 2023. https://www.census.gov/library/stories/2021/06/does-majoring-in-stem-lead-to-stem-job-after-graduation.html

20. Roberge, M. É., and Van Dick, R. "Recognizing the benefits of diversity: When and how does diversity increase group performance?" *Human Resource Management Review* 20.4 (2010): 295–308.

21. Rosenstein, A., Raghu, A., and Porter, L. 2020. "Identifying the prevalence of the Impostor Phenomenon among computer science students." In Proceedings of the 51st ACM Technical Symposium on Computer Science Education (SIGCSE '20). New York: Association for Computing Machinery, 30–36. https://doi.org/10.1145/3328778.3366815

22. Hutchins, H. M. "Outing the imposter: A study exploring imposter phenomenon among higher education faculty." *New Horizons in Adult Education and Human Resource Development* 27.2 (2015): 3–12.

23. Baumann, N., Faulk, C., Vanderlan, J., Chen, J., and Bhayani, R. K. "Small-group discussion sessions on Imposter Syndrome." MedEdPORTAL 16 (2020): 11004. https://doi.org/10.15766/mep_2374-8265.11004

24. "How we teach computing." National Center for Computing Education, teachcomputing.org/pedagogy. Accessed 23 July, 2023.

25. "7 research-based classroom strategies for teaching computer science." Engineering in Elementary, 17 May 2021, blog.eie.org/7-research-based-classroom-strategies-for-teaching-computer-science-cs. Accessed 23 July, 2023.

26. Wiggins, G., and McTighe, J. *Understanding by Design*. Association for Supervision and Curriculum Development, 2005.

27. McCarthy, J. "Learner interest matters: Strategies for empowering student choice." *Edutopia* (2014).

28. Bajorek, J. P. "Voice recognition still has significant race and gender biases." *Harvard Business Review* 10 (2019): 1–4.

29. McMillan, G. "It's not you, it's it: Voice recognition doesn't recognize women." Retrieved 4 February, 2011: 2017.

30. Computer scientist demographics and statistics [2023]: Number of Computer Scientists in the US. Zippia.com. Published 21 July, 2023. Accessed 23 October, 2023. https://www.zippia.com/computer-scientist-jobs/demographics/

31. Robison, M. "Voice assistants have a gender bias problem. What can we do about it?" (2020).

32. Henderson, S. "NIST study evaluates effects of race, age, sex on face recognition software." Text. NIST 19 (2019).

33. Ryan-Mosley, T. "The new lawsuit that shows facial recognition is officially a civil rights issue" (2021).

34. Brown, Lydia X. Z. et al. "Ableism and disability discrimination in new surveillance technologies." Center for Democracy & Technology (2022).

35. U.S. Equal Employment Opportunity Commission. "Diversity in tech" (2014).

36. Zippia. "Diversity in high tech statistics." Zippia, 7 November, 2022. https://www.zippia.com/advice/diversity-in-high-tech-statistics/. Accessed 7 November, 2022.

37. "Women in technology statistics." Techopedia. Accessed 15 July, 2023. https://www.techopedia.com/women-in-tech-statistics

38. Gruman, G. "IT snapshot: Ethnic diversity in the tech industry." Computerworld, 16 July 2020. https://www.computerworld.com/article/3567095/it-snapshot-ethnic-diversity-in-the-tech-industry.html. Galen Gruman. "IT snapshot: Ethnic diversity in the tech industry." Computerworld, 16 July 2020. https://www.computerworld.com/article/3567095/it-snapshot-ethnic-diversity-in-the-tech-industry.html

39. "Inclusive hiring practices in tech." TECHNATION. Accessed 15 July, 2023. https://technationcanada.ca/en/news/inclusive-hiring-practices-in-tech

40. Rock, D., and Grant, H. "Why diverse teams are smarter." *Harvard Business Review*, 4 November, 2016. https://hbr.org/2016/11/why-diverse-teams-are-smarter

41. "HR and recruiting stats." Glassdoor for Employers. Accessed 15 July, 2023. https://www.glassdoor.com/employers/resources/hr-and-recruiting-stats/

42. Scherer, R., Siddiq, F., and Sánchez Viveros, B. "The cognitive benefits of learning computer programming: A meta-analysis of transfer effects." *Journal of Educational Psychology* 111.5 (2019): 764.

43. Century, J., Ferris, K., and Zuo, H. "Finding time for computer science in the elementary day" (2017).

44. Trafton, A. "Cognitive scientists define critical period for learning language." MIT News, 1 May 2018. https://news.mit.edu/2018/cognitive-scientists-define-critical-period-learning-language-0501

45. The Education Hub. "6 strategies for promoting student self-efficacy in your teaching." The Education Hub, 2018. https://www.theeducationhub.org.nz/wp-content/uploads/2018/03/6-strategies-for-promoting-student-self-efficacy.pdf

46. Gladstone, J. R., and Cimpian, A. "Which role models are effective for which students? A systematic review and four recommendations for maximizing the effectiveness of role models in STEM." *International Journal of STEM Education* 8.1 (2021): 59.

47. Ha, S. et al. "Learning to walk in the real world with minimal human effort." arXiv preprint arXiv:2002.08550 (2020).

48. Brown, T. et al. "Language models are few-shot learners." *Advances in Neural Information Processing Systems* 33 (2020): 1877–1901.

49. Klee, M. "Professor flunks all his students after ChatGPT falsely claims it wrote their papers." *Rolling Stone*. Published 17 May, 2023. Accessed 23 October, 2023. https://www.rollingstone.com/culture/culture-features/texas-am-chatgpt-ai-professor-flunks-students-false-claims-1234736601/

Index

Solutions
YOU WANT

Experts
YOU TRUST

Results
YOU NEED

INSTITUTES

Corwin Institutes provide regional and virtual events where educators collaborate with peers and learn from industry experts. Prepare to be recharged and motivated!

corwin.com/institutes

ON-SITE PROFESSIONAL LEARNING

Corwin on-site PD is delivered through high-energy keynotes, practical workshops, and custom coaching services designed to support knowledge development and implementation.

www.corwin.com/pd

VIRTUAL PROFESSIONAL LEARNING

Our virtual PD combines live expert facilitation with the flexibility of anytime, anywhere professional learning. See the power of intentionally designed virtual PD.

www.corwin.com/virtualworkshops

CORWIN ONLINE

Online learning designed to engage, inform, challenge, and inspire. Our courses offer practical, classroom-focused instruction that will meet your continuing education needs and enhance your practice.

www.corwinonline.com

PLSN209A8

Visit www.corwin.com

CORWIN

A Sage Company

CORWIN HAS ONE MISSION: to enhance education through intentional professional learning.

We build long-term relationships with our authors, educators, clients, and associations who partner with us to develop and continuously improve the best evidence-based practices that establish and support lifelong learning.